Reprogen-ethics and the Future of Gender

INTERNATIONAL LIBRARY OF ETHICS, LAW, AND THE NEW MEDICINE

VOLUME 43

For further volumes:
http://www.springer.com/series/6224

Frida Simonstein
Editor

Reprogen-ethics and the Future of Gender

Springer

Editor
Frida Simonstein
Yezreel Valley College
D.N. Emek Yezreel 19300
Israel
fridas@yvc.ac.il

ISSN 1567-8008
ISBN 978-90-481-2474-9 e-ISBN 978-90-481-2475-6
DOI 10.1007/978-90-481-2475-6
Springer Dordrecht Heidelberg London New York

Library of Congress Control Number: 2009926865

Printed on acid-free paper

Springer is part of Springer Science+Business Media (www.springer.com)

Epigraph

New genetic technologies have implications for all fields in medicine, but when they are combined with reproductive technologies, the prospects are staggering. Indeed, the combination is so different from that of either technology alone that it deserves a new appellation: reprogenetics

(Lee Silver, Remaking Eden, 1998)

Not only are enhancements permissible but in some cases there is a moral duty to enhance ... enhancement is also an opportunity that it is in the interests of society and government to take ... parents would act ethically if they were to attempt to achieve such an objective for their children ...

(John Harris, Enhancing Evolution, 2007)

Acknowledgements

The idea of this collaborative volume matured in the course of summer 2007 during a Visiting Fellowship at the University of Liverpool at the Institute of Medicine Law and Bioethics, IMLAB. I am thankful to John Harrington at IMLAB who reviewed the first draft of Chapter 4. This project benefited from collected ideas in international meetings; particularly from John Harris (Beijing 2006), Matti Häyry (Barcelona 2005) and Daniel Callahan (The Hastings Center 2004). A travel grant of FAB allowed me to attend to the 2006 meetings of FAB and IAB in Beijing. Chapter 13 results from a Visiting Fellowship at the Hastings Center during summer 2004; it was first published in Philosophy of Medicine and Health Care (2006) 9: 359–365. Chapter 16 has a 'follow up' version to a discussion on the same matter printed in Cambridge Quarterly of Health Care Ethics (2008) 18: 1–8. Some of the ideas in the introduction appear in the Journal of Medical Ethics (2008) 34: 102–103. Chapter 12 was first published in Ethical Perspectives (2003) 10: 3–4.

Many thanks are due to the contributors of this volume; particularly to Daniel Callahan who provided useful advice; and Matti Häyry who suggested the papers by Tom Buller, Marleen Eijkholt and Iain Brassington.

Contents

Notes on the Contributors

Anna Smajdor is a lecturer in Ethics at the University of East Anglia, UK. She has a long-standing interest in the ethical aspects of science, medicine and technology, stemming from her first degree in philosophy. In 2004 she obtained 3 years' funding from the Wellcome Trust to investigate the ethics of using artificial gametes in reproduction. In 2007, she co-authored (with Baroness Ruth Deech) 'From IVF to Immortality' a book exploring legal and ethical controversies in reproductive technology and stem cell research. She is vice chair of the Riverside Research Ethics Committee, and an honorary research fellow of Imperial College, London, acsmajdor@googlemail.com

Ariel Revel MD is staff physician at the in vitro fertilization (IVF) unit and Department of Obstetrics and Gynecology, Hadassah Hospital, Jerusalem. He is Senior Fellow at the University of Oxford, UK, member of the Endocrinology circle, Hebrew University Medical School, manager of the oocyte donation (OD) program and leader of the fertility preservation program. He served as treasurer of the Israel Fertility Association (IFA) executive committee and is board member Special Interest Group for Endometriosis & Endometrium (SIGEE) of the European Society for Human Reproduction and Embryology (ESHRE), he is also at the Board of Directors for the cancer institute for fertility preservation, arielr2@hadassah.org.il

Daniel Callahan has a BA from Yale and a PhD in philosophy from Harvard. He was a cofounder of The Hastings Center and its President for many years. He is now Director of its International Program. His main focus over the years has been on biomedical ethics, but has also worked on matters of population and reproductive biology. In recent years he has turned his attention to health policy issues. He is an elected member of the Institute of Medicine, National Academy of Sciences and a Senior Lecturer at the Harvard Medical School. He is the author or editor of 40 books, most recently *Medicine and the Market: Equity v. Choice*, callahand@thehastingscenter.org

David Heyd is Chaim Perelman Professor of Philosophy at the Hebrew University in Jerusalem. His fields of research are ethics, political philosophy and bioethics. Among his publications are *Supererogation* (Cambridge 1982), *Genethics* (California 1992), *Toleration* (Princeton 1996), *Ethics and Medicine* (in Hebrew, 1989) and numerous articles. He has been active on various institutional and government committees in the sphere of medical ethics, including the National Council for Bioethics in Israel, david.heyd@huji.ac.il

Effy Vayena is Senior Research Fellow at the Institute for Biomedical Ethics at the University of Zurich. Prior to that, she was a staff member at the Department of Reproductive Health and Research in WHO where she was the responsible officer for issues related to infertility and assisted reproduction. She co edited WHO's decennial report on assisted reproduction (Current practices and controversies in assisted reproduction, WHO 2002) and coordinated several international activities on technical as well as ethical issues in this area. Her other areas of work include health research ethics with a focus on international clinical research as well as development of training materials for capacity building in research ethics, vayena@ethik.vzh.ch

Ekaterina Balabanova is a lecturer in Political Communication at the School of Politics and Communication Studies at the University of Liverpool, UK. Her research interests bring together the areas of international politics and international communication and include the relationship between news media and politics. Her research interests include media and human rights, particularly assisted reproduction, immigration and media power in Eastern Europe. Her book, *Media, Wars and Politics: Comparing the Incomparable in Western and Eastern Europe*, was published in 2007.

Ester Polak de Fried is the Chair of the Department of Reproductive Medicine at CER Medical Institute, Affiliated to the School of Medicine-Buenos Aires University (UBA), Argentina, pioneer in oocyte donation, oocyte cryopreservation and fertility preservation. She is Lecturer in Gynecology and Endocrinology at UBA and Vice-President of the Iberoamerican Eco-Bioethics Network, UNESCO Chair in Bioethics. She is a researcher at the Science and Technology Secretariat at UBA. She is board member of the Standards and Practice Committee of the International Federation of Fertility Societies (IFFS).

She became a permanent Advisor in New Reproductive Technologies, Women Health, Bioethics and Stem Cell Research at the Argentine National Congress. She is past President of the Argentinean Society for Reproductive Medicine and Past Treasurer of the IFFS. ester_polak@cermed.com, www.cermed.com.

Frida Simonstein is Senior Lecturer at the Department of Health Systems Management in the Yezreel Valley College and Research Associate at the Center for Women's Health and Promotion at the Ben Gurion University in Israel. She holds a PhD in Health Care Ethics from the Faculty of Law at the University of Manchester in the UK and a MSc in Human Genetics from the Sackler Faculty of Medicine at Tel Aviv University. She has been a Senior WHO Research Fellow and a Visiting Fellow at The Hastings Center. Her book *Self-evolution the Ethics of Redesigning Eden* was published by the International Center of Health Law and Ethics at Haifa University and Yozmot in 2004, fridas@yvc.ac.il

Harriet Bradley is Professor of Sociology at the University of Bristol, UK, and former Dean of the Faculty of Social Sciences and Law. She is a specialist in the study of women's employment; her broader research interests are the sociology of gender, feminism, industrial relations and the sociology of work, along with a general interest in social divisions and inequalities. Her publications include the books *Gender* (2007), Gender *and Power in the Workplace* (1999) and *Fractured Identities* (1996). She has recently completed a research project on workplace cultures and minority ethnic women for the UK Equal Opportunities Commission, Harriet.Bradley@bris@ac.uk

Iain Brassington is a lecturer in bioethics at the Centre for Social Ethics and Policy and Institute for Science, Ethics and Innovation at the University of Manchester. His research interests are wide, but major recent themes are self-killing and euthanasia, reproductive ethics, and autonomy and Kantian philosophy, iain.brassington@manchester.ac.uk

Marleen Eijkholt is an associate lecturer in the Center of Social Ethics and Policy at the University of Manchester. Her teaching and interests comprise medical law and bioethics. At the same time she works on her PhD about the right to procreate, within the program of Bioethics and Medical Jurisprudence. Her research has a legal-theoretical and comparative health law approach, marleen.eijkholt@manchester.ac.uk

Matti Häyry is Professor of Bioethics and Philosophy of Law at the University of Manchester in the United Kingdom. Before moving to Manchester in 2004, he held academic positions at the Universities of Helsinki and Kuopio in Finland, and Central Lancashire in the North West of England. His academic work has for the last two decades concentrated on moral theories and the ethics of genetics and reproduction. He has authored 12 and edited 13 books and volumes on bioethics and moral philosophy, and published c. 200 articles in peer-reviewed academic series and scholarly collections. He is also editor or board member in 9 international scientific journals and series in his field, including the *Journal of Medical Ethics*, *Bioethics*, *Theoretical Medicine and Bioethics*, and the *Cambridge Quarterly of Healthcare Ethics*. He is currently the President of the International Association of Bioethics, matti.hayry@manchester.ac.uk

Michal Mashiach-Eizenberg is lecturer in the Department of Health Systems Management at the Max Stern Academic College of Emek Yezreel. She has a B.A. and M.A in Statistics from the University of Haifa and a Ph.D. in Technological and Scientific Education from the Technion. Her teaching and research interests are in Statistics and Research Methods.

Søren Holm holds degrees in medicine, philosophy and religious studies, and health care ethics, as well as two doctorates in medical ethics. He is Professorial Fellow in Bioethics at Cardiff Law School and Professor of Medical Ethics at the University of Oslo. He has been researching issues in reproductive ethics since the mid 1980s, holms@cardiff.ac.uk

Tom Buller is Associate Professor and Chair in the Philosophy Department at the University of Alaska Anchorage. His main research and teaching interests are in bioethics and neuroethics, tombuller@uaa.alaska.edu

Tuija Takala is Adjunct Professor (Docent) in Practical Philosophy at the University of Helsinki, Finland. Tuija has written extensively on bioethical issues and her publications have appeared, for instance, in *Bioethics*; *Journal of Medical Ethics*; *Developing World Bioethics*, *Journal of Medicine and Philosophy*; *Medicine, Health Care and Philosophy*, *Medical Law International*, *Trames*, and *Theoretical Medicine and Bioethics*. She is Associate Editor of the *Cambridge Quarterly of Healthcare Ethics* and President-Elect of the European Society for the Philosophy of Medicine and Health Care. Tuija's research interests include the conceptual and theoretical dimensions of bioethics and recently she has been interested in questions related to personal identity, tuija.takala@helsinki.fi

Introduction

Frida Simonstein

The origin of the present volume can be traced back to the 2006 Congress of the International Association of Bioethics in Beijing. At that meeting, there was a special session on human enhancement in which panelists addressed important issues, such as biopsychological enhancements. The possibility of regulating emotions through pharmacological means, biases that may affect our judgments against human enhancement, health care inequalities that will follow from the adoption of genetic technology, the social impact and cost if the new technology is accepted, and women's equality by genetically becoming as strong as men were all discussed [1]. John Harris, who has recently claimed that there are ethical imperatives that will inevitably lead to the further deliberately chosen evolution of the human species [2] concluded is resumed the right word? that session. None of the participants was interested in addressing the woman's role in such a scenario; which is not necessarily surprising as one session alone cannot discuss all the issues related to a subject. Yet *none* of the defendants of enhancement, up to the time of writing these lines, have ever addressed women's crucial involvement in enhancing. Moreover, Harris' claim above seems to imply that it is women's *dut*y to enhance future generations. This implies that all women would be *required* to reproduce only by means of IVF. Nevertheless, despite the far reaching implications for women, Harris does not mention their role.

It is the purpose of this volume, therefore, to offer a comprehensive platform of discussion on the main and neglected factor in the equation of enhancing, namely its 'vessel'. The book presents topics that are often discussed separately in bioethics, but are totally interrelated: assisted reproduction, enhancing and gender.

Reprogenetics and Enhancing

Reprogenetics refers to the merging of reproductive and genetic technologies [3]. These can prevent or ensure the inheritance of particular genes in a child. According to Lee Silver, the main motivation of reprogenetics is 'the desire of parents to give all possible advantages to their children' [4]. While reprogenetics may avoid having a child with severe impairment, it may also pave the way for genetically enhanced future generations.

F. Simonstein (ed.), *Reprogen-ethics and the Future of Gender*, International Library of Ethics, Law, and the New Medicine 43, DOI 10.1007/978-90-481-2475-6_1,

Enhancing may be necessary and even morally required [5], but it remains highly controversial [6]. Some forms of enhancing, however, are more acceptable than others. For instance, genetic screening for inherited diseases and malformations, which is an indirect form of enhancing, is widely accepted. Screening for these diseases has become common in the developed world [7]. However, as more information becomes available, more tests are offered for rarer genetic diseases, some of which are milder and/or treatable. As a result, there is a growing package of genetic tests that are becoming normative in some circles. In Israel, 94.4% of secular and educated women take a wider package of genetic tests [8].

Genetic Screening: Are Women Worse Off?

Some people claim that too many genetic tests affect women badly. Larisa Remenick has studied the reasons why women pursue genetic screening in Israel and she argues that the availability of genetic screening for rarer and/or treatable diseases makes Israeli women worse off [9]. Yet, Remenick's claim does not take into account that those who will have to shoulder the consequences of not testing (by looking after a chronically ill child) would most probably be the women, even with the best supportive health care available (which is not often the case).

In her study, women articulately explain that they want to keep 'out of trouble' by avoiding giving birth to a child with disease, even if this means stopping a pregnancy [10]. Thus women's attitudes are practical and 'prophylactic'. While many people may disagree with and/or dislike such reasoning, this approach does not necessarily make women worse off. Women also observe that men care less about genetic testing because the burden of care, in the event that a child happens to be born with a disease, would fall mostly on the women. Thus, widening the package of genetic tests seems to be helping Israeli women to make decisions based on their own needs and best interests. This is in contrast to Remennick's conclusions.

Other Ways of Enhancing: PGD and IVF

Other ways of enhancing, however, are a very different story.

Enhancing either by choosing an embryo through pre-implantation genetic diagnosis (PGD) or by genetically enhancing gametes or embryos (when and if it becomes safe) could effectively protect future generations from ill health [11]. However, this way of enhancing will have to use techniques of *in vitro* fertilisation (IVF) in order to choose and/or implant an enhanced embryo (see Chapter 1). This is quite different from testing for genetic disease, which requires only a blood sample in most cases. IVF is an invasive procedure, unhealthy, inefficient, demanding and extremely unpleasant for the women involved [12]. This way of enhancing could hardly be in the women's best interest since a not-enhanced child could be reasonably healthy [13].

Women's role in this scenario of enhancing has been overlooked. This may be because women's involvement as reproducers is taken for granted; and/or perhaps because already *millions* of women go through IVF in order to overcome infertility. Certainly, assisted reproduction (ART) by means of IVF has become a routine procedure in many parts of the world (see Chapters 1, 2, 4). ART has also become a matter of human rights (see Chapter 3) and a topical issue in both developed and developing countries (see Chapters 4, 5, 6).

Nevertheless, IVF is an extreme solution, pursued because it is a woman's last chance to conceive a child. IVF is a well-known nightmare [14]. Thus a requirement of 'universal' reproduction by means of IVF in order to enhance the human race would significantly add to women's reproductive burdens. Therefore women's 'expected' role in this scenario should be clearly spelled out by the defenders of enhancing.

Reproductive Choice?

At this point many people would claim that enhancing would be a woman's 'choice'. Certainly there is an extensive body of feminist literature about women's choices on reproductive technologies. During the seventies of the 20th century Shulamith Firestone wrote that technology can liberate women from their biology and extend women's choices [15]; but other feminists claimed that the technological 'invasion' of women's bodies by science amounts to 'a patriarchal strategy to take away from women the one advantage they have over men' [16]. This claim has been regarded as overstated by other feminists who have argued that such interpretation places too much emphasis on 'patriarchal' medicine and science while minimizing the knowledge and agency of women themselves [17]. Feminists have also suggested the notion of 'relational autonomy' an idea that takes into account the social 'embeddedness' of women's reproductive choices [18, 19].

In this volume, Søren Holm and Harriet Bradley point out (in Chapters 2 and 9 respectively), that since the advent of IVF, reproduction has been increasingly medicalised, affecting the freedom of would-be parents. Holm and Bradley both argue that the increase in 'procreative liberties' created by the developments of ART has paradoxically decreased the effective freedom of some people. According to Holm, this situation has emerged because the broad social acceptance of ART in many countries has changed social default expectations about reproductive choices. Thus, anyone who admits to being infertile yet still has a desire for children is likely to be met with the expectation that ART treatment is there for the taking (see also Chapter 4) [20].

Certainly, as Anna Smajdor reminds in Chapter 8, any pregnancy is intrinsically risky. Yet, as she observes, this is a *necessary* risk and thus highly encouraged by society (Chapter 8). Genetically enhancing future generations by means of IVF would be even more dangerous for women; but paradoxically, enhancing could be *medically* prescribed. As a result, enhancing could also become largely encouraged by society.

Supposing that genetic enhancing was established, could a woman *really choose* to refuse granting her child a healthier outcome? Hardly. It could be argued that enhancing her child would be any mother's expectation. This may be true; but if so, having to go through IVF to achieve this would not quite be a *real* 'choice'. Enhancing or not would be a conflictive situation for a woman; but because motherhood is highly idealized as the altruistic and selfless enterprise that characterizes womanhood, women seemingly are bound *a priori* to accept. Moreover, as the 'ideal' mother is socially determined (see Chapters 4, 5, 6, 7), it is not implausible that a woman's refusal to enhance her child could become to be seen as 'selfish' and therefore socially condemned [21].

Education

Probably the next argument would be that women's (in)ability to hold autonomous choices in reproductive matters is unrelated to the ethics of enhancing. As Søren Holm points out, this is a common argumentative move in bioethics: if a certain negative effect of an act or policy can be traced to some kind of prejudice or ideology in society, then that is not a reason not to perform the act or introduce the policy, but a reason to eradicate the prejudice or ideology. Yet as Holm rightly observes, 'it is seldom explicated how such an eradication should be achieved, or who might be responsible for achieving it' (see Chapter 2).

Indeed, education is a key for eradicating prejudices and ideologies. Yet, how could the ideology of motherhood be challenged? Conditioning towards motherhood (preferably unconditional and altruistic) as equivalent to womanhood remains built-into the early socialization and later education of girls (see Chapters 4, 5, 13). In many societies childless women are still considered 'defective' (see Chapters 4, 5, 8). Although for a given woman remaining childless may be healthier, voluntary childlessness is very much frowned upon.

In some societies voluntary childlessness is socially castigated and therefore inexistent or hidden. In Israel, women who are voluntarily childless would prefer to say that they have some infertility trouble rather than openly admit that they do not want to have children. Even well-educated women in Israel think it is their *duty* to have children (see Chapter 4). Necessary motherhood remains ingrained, particularly in pronatalist societies; however, this is also the case in both developing and developed countries (see Chapters 4, 5, 14).

Choice Revisited

In a post-feminist era, in which the notion of 'choice' has become a key term (at least in western societies), many women *must* recognize, perhaps unwillingly, that reproduction remains pretty much a predetermined activity. As only women can bear children [22] many women must *necessarily* adjust to the fact that they are the reproductive channel of any given couple. As the reproducers of the human species, women still have more duties than choices; these duties may include the wishes of the partners, the hopes of the family and/or the expectations of society (see Chapters 4, 5, 7, 8, 9) [23]. Unless a woman chooses to remain childless (if she can) it is the woman who is *inevitably* required to endure pregnancy and childbirth. Indeed, Harris' moral duty to enhance future generations may become just an additional demand.

Certainly not all women feel oppressed as reproducers. People may also point to the flourishing industry of assisted reproduction and its images of 'desperate' women who cannot conceive a child. Yet the fact that women may want to have children – and it is irrelevant here that it is impossible to distinguish whether women's 'despair' comes from hormones or conditioning – does not necessarily mean that they are less constricted by their biology.

My personal experience is that I wanted to have a family and I am the proud mother of a daughter and a son. However, I could have been equally a happy and proud parent if my partner had gone through childbearing and childbirth. As we both wanted to have children, but my partner could not bear them, I *had* to do the job for both. Of course, men may also feel constricted by their biology as they cannot be the bearers of their offspring. However, unlike women, men can still enjoy parenthood without having to endure the burdens and risks involved in childbearing and childbirth; a point to which I will return.

The Future of Gender

> *[W]omen will all too ruefully understand that men are still men, for better or worse. I won't say we men can't help ourselves, but there is considerable biology and ingrained acculturation to overcome, and I expect there always will be. As a veteran of the androgyny bubble (circa 1960/1970)—the utopian belief that gender is nothing more than a repressive social construct, soon to be transcended—there is nothing like raising actual boys and girls to get over that fantasy* (Dan Callahan, Chapter 7).

Some people may opt to remain childless. And yet, while today assisted reproduction is forwarded in terms of 'rights', reproduction is also regarded as an individual 'duty' to society. Thus people are expected to have children; and those who do not have them (for any reason) are socially required to support (at the least) the reproductive enterprise of others. In Chapter 10 Matti Häyry challenges these premises. Häyry asserts that he does not find good grounds for the view that humanity as a whole should continue to exist; neither minds a 'peaceful demise of the human race' (see Chapter 10). Häyry's view seems atypical; certainly from the point of view of a pronatalist country from which I am writing these lines [24]. Yet in Europe reproduction has plummeted even below the replacement rate. Indeed, Häyry's ideas of voluntary childlessness seem to be less unpopular in some parts of the world. While seems unlikely that humans will give up reproducing in a near future, the idea of the continuity of the human race is quite at the centre of gendered roles. Thus Matti Häyry's notions of non-parenthood could deeply affect future perceptions of gender.

Will equality of gender remain a utopian fantasy, as Callahan believes? This is difficult to predict. With the introduction of extracorporeal gestation by IVF, it was suggested that gender equality could be achieved with the aid of artificial wombs and ectogenesis, a child brought to term outside a woman's womb [25]. Tuija Takala reflects on this idea in Chapter 14. The artificial womb may indeed arrive earlier than we think, as different areas of science and technology are converging towards the knowledge of ectogenesis (see Chapter 13). However, ectogenesis is not exactly welcome. While this development could allow women to have *real* reproductive choice and also enable enhancement without necessarily having to enroll all women in IVF, our present intuitions and legislation are very much at odds with such development (see Chapters 15 and 16).

Reading this Book

As Daniel Callahan observes in Chapter 7, most books and articles about gender have been written by women and for women. Indeed, discussions on reproductive and/or genetic technologies which involve the dimension of gender are often seen as items undertaken by and only for 'feminists'; thus either to be engaged with, or to be avoided. This volume was not aimed at a feminist discussion of reprogenetics; as it aimed to be inclusive rather than exclusive to include in the discussion authors whose ideas and/or work may affect gender but whose orientation is not 'feminist' necessarily.

The volume presents data, analyses and lines of thought about the future of human reproduction that have emerged during the first decade of the 21st century projecting towards the future. Chapters 12, 13 and 16 have had earlier versions and have been included in this book to provide the reader with a fuller perspective. As it aims at reaching a multidisciplinary audience, the book is multi and interdisciplinary in scope with contributors from the fields of medicine, genetics, law, philosophy and sociology. This book does not provide definite answers, but offers a wider perspective of the complex issues involved. Its main purpose is to set a framework for further discussion.

Part I introduces the reader to a 30-year retrospective of assisted reproduction (Chapters 1, 2, 3). This section presents the actual state of the ARTs; a reflection from a 30-year retrospective; and a discussion of the right to procreate. Part II focuses on the lack of ART in less developed countries; but it also provides a paradoxical perspective from a developed country (Chapters 4, 5, 6). Some of these discussions may not address gender directly. Yet the gendered dimensions of the first two sections of this book stem from the fact that IVF has not yet been tried on men. As the medical enterprise which set reprogenetics off, IVF cannot be ignored in the context of this book. Part III points more directly to the gendered dimensions of ART, some of its actual effects on women's choices and some of the societal expectations from women (Chapters 7, 8, 9). Part IV looks at some parental (and non-parental) perspectives emerging from the first decade of the 21st century. The first chapter in this section challenges traditional notions such as the assumption of universal willingness to parent a child and the idea of societal solidarity on reproductive matters. While this discussion does not address assisted reproduction, it challenges assumptions that have been at the center of the IVF industry and the appeals for the funding of IVF by the state(s). The essays in this section provide also ethical analyses of parental reprogenetic choices which may likewise apply for enhancing (Chapters 10, 11, 12). Finally, part V does project from present developments into the future. This section deals with some of the perspectives posed by the advent of the artificial womb; and particularly aims at encouraging further discussion. A pilot study of people's views on this development closes this section (Chapters 13, 14, 15, 16).

Detailed Outlines

In Chapter 1, Ariel Revel provides a detailed account of the scientific state of assisted reproductive techniques (ART). Revel surveys the various advancements in reproductive medicine, which, in his words, 'have now gone down to the very basic building blocks of biology.' He observes that the field of preimplantation genetic diagnosis (PGD) is constantly progressing and even refers to the ability to derive embryonic stem cells from human embryos for research. He concludes that while innovative technologies employed in reproductive medicine may offer humankind many hopes, they are also the source of need for careful considerations and observation.

In Chapter 2, Søren Holm observes that almost 30 years after the birth of Louise Brown it is relevant to assess how ART has affected reproductive practices and our view of reproduction. He asserts that ART has technologised, medicalised and commercialised reproduction. Holm argues that while ART seems to have led to a *prima facie* state of reproductive freedom where the consumer is king (or more accurately queen) the medical profession still shapes ART practices. Holm believes that ART continues to contribute to a medicalization of reproduction that constrains the space of action for would-be reproducers.

In Chapter 3, Marleen Eijkholt claims that while some authors argue that the moral relevance of the right to procreate can be equated with the right to rear children, this emphasis can be misleading. She argues that the moral and legal relevance of the right to procreate lies in the *capacity* to procreate, or the capacity to conceive offspring. Eijkholt believes that placing the focus on a right to realise and control one's biological capacity to procreate makes it possible to create a coherent moral and legal human right.

In Chapter 4 Ekaterina Balabanova and Frida Simonstein present a comparative review of ART's provision in two pro-natalist countries. The chapter examines women's status in Bulgaria and Israel and both states' policies and involvement in assisted reproduction. The focus is on assisted reproduction as reflected by present policies in both countries and how this influences women's reproductive choices, and whether assisted reproduction in these countries makes women better off.

In Chapter 5, Effy Vayena emphasizes the global disparity in assisted reproduction. Vayena points out that in developing countries infertility is a low priority area of health policy despite its relatively high prevalence and the cultural values associated with childbearing. She challenges arguments against reproductive technologies in the developing world and discusses the needs of infertile couples in the context of reproductive autonomy and the right to the highest attainable standard of health.

In Chapter 6, Ester Polak de Fried analyses the reasons for unequal access to fertility treatments and the way in which the different policies of many Latin American countries affect this situation. Polak de Fried presents a summary of the ethical aspects that influences ART in Latin America from the social, economic and individual point of view. She emphasizes the necessity of equal access to preventive medicine, to the health care systems and to ART procedures in Latin America.

In Chapter 7, Daniel Callahan observes that the conventional wisdom is that women with professional aspirations should gain their education and professional credentials before having children as a matter of commonsense. However, he points out that postponing having children has generated some unfortunate consequences, both professionally and maternally. Callahan proposes that women should be encouraged to reproduce at an earlier age, in order to avoid the risks associated with advanced maternal age.

In Chapter 8, Anna Smajdor questions Callahan's proposal. She observes that in Britain two spectres of 'inappropriate' reproductive choice have been increasingly prevalent in the media: either the young mother, feckless, ignorant and dependent on state handouts for survival; or the selfish older woman attempting to manipulate her biological functions to fit her career schedule. Smajdor argues that what constitutes a 'biologically optimal' time for reproduction is largely socially determined.

In Chapter 9, Harriet Bradley examines the future impact of reproduction needs and choices on women's lives and gender relations. Bradley argues that there is a paradox: technologies of reproduction, both established forms of contraception and new forms of fertility control, have maximised choices for women, offering them far greater control about whether, when, with whom and how they have children. Yet, she observes, anxieties around fertility and reproduction do not seem to have diminished, while childbearing and childcare responsibilities are still preventing women from attaining equality with men in the public sphere.

In Chapter 10, Matty Häyry discusses possible responsibilities towards future generations. He observes that there are two questions that precede any considerations of reproductive genetics and gender. They are 'Should people have children to begin with?' and 'Who is responsible for children and why?' Following the line of thought of his previous publications Häyry contends that if it is irrational or immoral to produce offspring, reproductive genetics turns out to be a waste of time and gendered family roles obsolete.

In Chapter 11, Tom Buller in response to the previous chapter, examines the rationality of not having children to avoid the risk that the future child will have a life of negative value. Buller argues that the decision facing prospective parents is not restricted to a choice between a negative and a neutral outcome and that it is impossible to avoid gambling on our children's lives. He claims that despite the fact that a future life cannot logically have value, it is still seen as superior in comparison to a wrongful life.

In Chapter 12, David Heyd analyses arguments for and against the practice of sex selection for non-medical reasons using PGD. Heyd distinguishes between arguments about the risks to the future child, the mother and society, and the inherent wrongness of the practice as an illegitimate interference in the natural course of reproduction. Heyd argues that despite the intuitive objection to the practice of sex selection, it can be justified in terms of parental autonomy and the value of family planning.

Chapter 13 presents research that is leading towards the advent of the artificial womb and ectogenesis. An aggressive and socially encouraged medical approach is observed, which saves fetuses in very premature deliveries and seeks to improve assisted reproduction by means of IVF. Given that prematurely born babies spend notably less time in a woman's womb, this begs the question if babies should necessarily be inside a woman's body for any period.

In Chapter 14 Tuija Takala analyses the philosophical, moral, social and political problems that arise from the present division between the two genders before advancing to refer to a possible future world where ectogenesis is normal. Takala looks at the ways in which such a world would undermine the currently prevailing gender roles.

In Chapter 15, Iain Brassington observes that parenthood is regarded as a joint project in which people are parents, irrespective of whether or not the child has been born. He notes that because of this, some people could suppose that fathers 'ought' to have a say in decisions concerning abortion. A strong reason to reject this would be the fact that the fetus is attached to the mother, but Brassington notes that ectogenesis may alter this picture because the status of the parents is equalised.

Chapter 16 presents a pilot study of people's opinion in Israel on the development of the artificial womb (AW). This study, focusing on the acceptability of this development, was conducted with a structured self-report questionnaire. Most participants in this study considered acceptable the use of AWs for some reproductive purposes, but not as a means to alleviate women's reproductive burdens. Only ten percent of the respondents were willing to use AWs to avoid a pregnancy; but this could amount to a considerable number of people in a given population.

References

1. The 8th World Congress of Bioethics. 2006. Beijing, China: Book of Abstracts, pp. 138–9.
2. Harris, J. 2007. *Enhancing Evolution*. Princeton, NJ: Princeton University Press.
3. Silver, L.M. 1998. *Remaking Eden: How Genetic Engineering and Cloning will Transform the American Family*. New York: Avon Books.
4. Silver, L.M. 2000. Reprogenetics: third millennium speculation The consequences for humanity when reproductive biology and genetics are combined. *EMBO Reports* 1(5), 375–378.

5. I have agued elsewhere, that enhancing future generations may be necessary because health care systems in the developed world are nearly collapsing under the pressure of late onset chronic diseases, Simonstein F. 2004, *Self-Evolution. The Ethics of Improving Eden*. Tel Aviv: Yozmot. Largely successful in preventing early deaths in the last century, health care systems have prompted a new plague, recognized lately by the World Health Organization as a new pandemic of noncommunicable, chronic diseases. See, Innovative Care for Chronic Conditions. 2002, Global Report. World Health Organization. While bad habits may result in bad health, some people who live abusive life styles may not suffer the ill consequences of others, who may squarely observe the golden rules of good health. This is because there is a genetic *predisposition* to cancer, diabetes, Alzheimer and cardiovascular diseases. In short, observing healthy life styles may *delay* the onset of chronic disease in persons who are at risk; which is important; however, this would not prevent the onset of the disease, eventually. Some WHO reports do not ignore that there is a genetic component to these diseases, Genomics and World Health, 2002 World Health Organization.
6. Fukuyama, F. 2002. *Our Posthuman Future*. London: Profile Books.; Shakespeare, T. 1995, Back to the future? 'New genetics and disabled people'. *Critical Social Policy* 44, 22–35; *Reproduction and Responsibility. A report of the President's Council on Bioethics*. 2004 Washington, D.C. Hardcopy. Also at www.bioethics.gov. This list is not exhaustive. The WHO has therefore chosen to focus on the uncontroversial change of life styles. See, Care for Chronic Conditions. 2002, Global Report. World Health Organization.
7. Even some developing countries have introduced genetic screening programs for some genetic threats which affect particularly their populations. Medical genetic Services in Developing Countries. 2006. World Health Organization.
8. Sher, C. et al. 2003. factors affecting performance of prenatal testing by Israeli Jewish women. *American Journal of Medical Genetics* 120A, 418–422.
9. Remennick, L. 2006. The quest for the perfect baby: why do Israeli women seek prenatal testing? *Sociology of Health & Illness* 28, 21–53.
10. Moreover, women's determination to take these genetic tests is consistent with a WHO report on genetic testing, which remarks that women "should be making decisions about genetic testing on the basis of their own needs and interests."WHO Report (Genomics and World Health) 2002, pp. 160.
11. Simonstein, F. 2004. Germ line engineering (GLE) and late onset diseases. The ethics of self evolution *Israel Medical Association Journal* 6, 652–657.
12. Remennick, L. 2000. Childless in the land of imperative motherhood: Stigma and coping among infertile Israeli women. *Sex Roles* 43, 821–884.
13. The rate of success for IVF is very low. Kaplan, E.H., A. Hershlag, A. H. DeCherney and G. Lavy. 1992. To be or not to be? That is conception! Managing in vitro fertilization programs. *Management Science* 38, 1217–1229. In Israel the official publications of IVF success is 30%; but this rate refer to 'pregnancies' and not to 'take home babies': in 2003 the rate of 'take home baby' was 14%. Simonstein F. 2006 Pressures on women to reproduce and the drive towards assisted reproductive technologies. *Med Law* 25, 355–363.
14. Remennick, L. 2000. Childless in the land of imperative motherhood: Stigma and coping among infertile Israeli women. *Sex Roles* 43, 821–841.
15. Firestone S. 2003 [1970]. *The Dialectic of Sex*. New York: Farrar, Straus and Giroux.
16. Corea et al. 1987. *Man-made Women*. Bloomington: Indiana University Press; Arditti et al. 1989 *Test-tube Women* London: Pandora Press. Cited in Pilcher J. and Whelehan I. 2005 (third edition) *50 Key Concepts in Gender Studies*. London: Sage; pp. 142.
17. Steinberg, D. 1997. *Bodies in Glass: Genetics Eugenics and Embryo Ethics*. Manchester: Manchester University Press.
18. Sherwin, S. 1998. 'A Relational Approach to Autonomy in Health Care.' In *The Politics of Women's Health: Exploring Agency and Autonomy in Health Care*. Ed. Susan, Sherwin Philadelphia: Temple University Press.
19. MacKenzie, C. and Stoljar, N. 2000. 'Introduction: Autonomy Refigured.' In *Relational Autonomy: Feminist Perspectives on Autonomy, Agency, and the Social Self*. Eds. Catriona MacKenzie and Natalie Stoljar. Oxford: Oxford University Press.

20. This pattern can be recognized in Israel, where, exceptionally, even the monetary obstruction to pursue ART is inexistent. IVF in Israel is offered to all women who are Israeli citizens, free of charge up to two children (see Chapter 4). As Rickie Solinger points out: "intensely private decisions about reproduction, including decisions about getting pregnant or not, staying pregnant or not, being the mother to the child one gives birth or not, are always shaped by public laws and policies. This may be a particularly difficult insight to bring into focus, in part because of the way 'personal choice' has eclipsed all other ways of thinking about pregnancy and motherhood." Solinger, R. 2005. *Pregnancy and Power*. New York: New York University Press.
21. Moreover, since it would be a refusal to a *medical* prescription, the rejection of enhancing could even end up being viewed as 'parental negligence' and therefore, punished by legislation.
22. During 2008 a man gave birth to a child; but this could happen only because he was female-borne, and still had a womb.
23. Women's choice in reproductive matters may have become somewhat more manageable since the advent of the pill; yet unwanted pregnancies are not uncommon. This is an awful situation for women even having accsess to safe abortion (which is still forbidden today in 80 percent of all countries).
24. In Israel cycles of IVF up to two children per israeli citizen women are paid by public funds (see Chapter 4).
25. Singer, P. and Wells, D. 1984. *The Reproduction Revolution. New ways to Making Babies.* Oxford: Oxford University Press; Firestone, S. *The Dialectic of Sex: The Case for Feminist Revolution* (1970. New York: Farrar, Straus and Giroux, 3rd ed. 2003).

Part I
ART – A 30 Years Retrospective

Chapter 1
Current Status of Assisted Reproductive Techniques (ART)

Ariel Revel

Introduction

Three decades have passed since the birth of Louise Brown, the first child born as a result of in vitro fertilization (IVF) [1] and it is difficult to imagine the world without assisted reproductive technologies (ART). The pioneering research leading to that development seemed cutting edge at the time [2]. The first pregnancy obtained by IVF implanted in the tube demonstrating one of the many risks entailed in separation of biological and social parenthood. The debate surrounding the ethical implications of ART as well as the cost and complexity of treatment has been a part of this technology ever since [4, 5]. Despite much initial resistance by the medical community and by society, IVF has now firmly established its place in the clinical management of infertility. And just around the corner future technologies are evolving including oral as well as sustained-released gonadotropins administration that will facilitate their intake and somewhat decrease the therapy burden from women. Innovative technologies employed in reproductive medicine in the 21st century offer many positive prospects while requiring extensive observation and debate on ethical issues. This chapter details some of the various advancements in reproductive medicine, which go into the very basic building blocks of life.

Indications of ART

Whereas IVF was originally devised to circumvent mechanical infertility, it soon became clear that ART could solve the majority of infertility problems indication including mild male infertility [6], severe male infertility requiring gamete micromanipulation, polycystic ovarian syndrome, endometriosis, poor ovarian reserve and unexplained infertility. Conventional treatment of the infertile couple has been accessed by single would-be parents and same sex couples asking for ART to help

A. Revel (✉)
Department of Obstetrics and Gynecology, Hebrew University-Hadassah Medical School, Jerusalem, Israel
e-mail: arielr2@hadassah.org.il

F. Simonstein (ed.), *Reprogen-ethics and the Future of Gender*, International Library of Ethics, Law, and the New Medicine 43, DOI 10.1007/978-90-481-2475-6_2,

them achieve parenthood. Within a lesbian relationship ART simply requires donor sperm, one female partner to provide the oocyte and the other female partner to act as embryo recipient. More recently homosexual men have been showing interest in fatherhood [7]. The technique would mandate an egg donor and a gestational surrogate [8]. If society finds these applications to be acceptable, surrogacy and using donor egg and sperm would be much in demand to provide families for same sex couples [9].

The trend towards delaying the birth of a first child until an age at which female fecundity or reproductive capacity has been lowered has increased the incidence of age-related infertility. In about half of the IVF cycles reported in Europe and America the woman is 35 years of age or older, reflecting this demographic trend [10, 11]. Unfortunately, IVF by itself cannot compensate for the lower fecundity associated with increasing age [12]. Expectations from IVF should therefore be realistic. Women should realize that although IVF is a very strong tool, its efficacy drops after age 35 and plummets after age 40.

Ovulation Induction

Controlled ovarian stimulation (COS) induces the superovulation required by ART to obtain several oocytes for fertilization. Tailoring ovarian stimulation to the individual patient is challenging, as ovarian response is considerably variable. This individual variability in ovarian response to COS necessitates dose adjustment by strict patient monitoring. Whereas exaggerated ovarian response might result in ovarian hyperstimulation syndrome [OHSS], obtaining too few oocytes may decrease pregnancy rates. Pharmacogenetics has emerged as a new area of research to improve the balance between desired and undesired actions of drugs, based upon the genetic predisposition of the individual patient. Genetic factors may thus eventually help predict ovarian response and the likelihood of OHSS. Clinical studies have demonstrated that FSH-receptor gene polymorphism determines the ovarian response to FSH stimulation in patients undergoing IVF. Genotyping patients scheduled for ovarian stimulation may soon individualize FSH dosing according to genetic differences in ovarian sensitivity [13]. The current practice of individualized treatment is based only on clinical experience and has poor predictability. Another challenge is to develop prediction models for treatment response in ovulation induction as the first steps towards evidence-based individualized treatment [14].

Ovarian reserve reflects the reproductive age of an individual woman and refers to what remains of the ever-declining pool of primordial follicles in the ovaries at a given time and the reproductive potential of each oocyte. Markers of ovarian reserve such as antral follicle count (AFC) and measurement of the anti- mullerian hormone (AMH) are applied in many IVF units to predict ovarian response. Although ovarian reserve is likely to be linked to the clinical response to exogenous stimulation it is unclear whether there exists a distinct threshold. Moreover, the likelihood of pregnancy in a woman undergoing COS is subject to a large number of factors other than ovarian reserve and ovarian response.

Though use of urinary gonadotropins was described 45 years ago [15] the new commonly used costly recombinant FSH compounds are finding it difficult to prove much superior efficacy [16]. Newer products that combine FSH and LH and the future possibility of one-dose injections or even oral gonadotropins may remove the need for daily injections.

Severe Male Infertility and Intracytoplasmic Sperm Injection [ICSI]

No effective treatment for severe male factor infertility was available until 1992, when intracytoplasmic sperm injection (ICSI) was introduced as part of the IVF process [17]. Considering the remarkable success of ICSI, many fertility centers have extended the indications for ICSI even when semen measurements are normal to include cases of unexplained infertility, fertilization failure in a previous IVF cycle, few or poor-quality oocytes available for insemination and gamete donation. Some centers have even advocated for the routine use of ICSI in all IVF cycles. Proponents of such indications for ICSI suggest that its routine use generates more accurate information about oocyte quality and maturity, bypasses any potential barriers to fertilization, and optimizes fertilization outcomes [18]. This trend is surprising, however, since several studies have shown that clinical outcomes are not improved with ICSI for infertility indications that are not attributable to male-factor conditions [19]. Though a 10 year follow-up of children born after ICSI is reassuring [20], the effects of ICSI on imprinted genes and genes involved in DNA replication error repair have raised some concern [21]. Sperm retrieval directly from the testes by extraction biopsy or fine-needle aspiration enables to offer fertility treatment to azoospermic men [22, 23].

Gamete Donation

Sperm and egg donation achieve high fertilization rates in IVF treatments mostly because they are obtained from young, healthy donors following screening. The guidelines for Gamete and Embryo Donation [24] recommend evaluation of potential sperm, oocyte, and embryo donors for sexually transmitted infections, genetic diseases, and psychological stability. The rising demand and increasing scarcity of donor oocytes in the USA has resulted in very high reimbursement rates for donors and to some fertility clinics sourcing oocytes donors from abroad, particularly from poorer countries, in what is referred to as 'international' oocyte donation [25, 26]. The large demand and the significant legal and ethical differences between European countries for egg and sperm donation have led to the movement of infertile patients across legislative borders to seek a solution to their problems [27]. Although currently there are no two European countries with the same regulations, a recent directive of the European parliament [28], which covers all banked tissues used in therapy including gametes, embryos and stem cells, has had a growing effect on how sperm bank and probable future egg banks will function.

Reproductive tourism is the traveling by candidate service recipients from one institution, jurisdiction, or country where treatment is not available to another institution, jurisdiction, or country where they can obtain the kind of medically assisted reproduction they desire. Patients travel for egg donation, preimplantation genetic diagnosis for sex selection for non-medical reasons, insemination with sperm from a relative or homosexuals who find surrogate mothers for their embryos. It can be predicted that this type of traveling will steadily increase. There are indications that most patients are prepared to go abroad to get the type of treatment they want.

Anonymity of the donor is a very controversial issue. In 2005, UK legislation was changed requiring any donor of gametes or embryos to agree to the disclosure of his identity to any offspring reaching the age of 18. Removal of anonymity for gamete donors has led to further restriction of an already insufficient supply to patients requiring donated eggs [29]. The use of double donation to single women raises medical and ethical questions, though the welfare of the child does not seem in jeopardy [30].

In Vitro Maturation of Oocytes

Immature human oocytes obtained from patients can be matured and fertilized in vitro (Cha and Chian, 1998). Human oocyte maturation is considered as the reinitiation and completion of the first meiotic division from the germinal vesicle stage (prophase I) to metaphase II, and the accompanying cytoplasmic maturation for fertilization and early embryonic development. This technique is mostly suitable for polycystic ovary syndrome (PCOS) patients a high risk for OHSS, as it avoids the risks and side effects of ovarian stimulation. Although there has been significant progress in terms of improving clinical pregnancy and live-birth rates after IVM, there remains much to learn in terms of what controls maturation, fertilization, and implantation rates. Compared with IVF and ICSI, IVM is not associated with any additional risk [31, 32].

Culture Media

Most embryo culture media used in human IVF are based upon media that were successfully used in animal embryo culture [33]. Since the development of Human Tubal Fluid Medium (HTF), specifically designed for human IVF, many modifications and advancements have been made in embryo culture medium formulae. Sequential media, developed in order to respond to the changing metabolic needs of the rapidly growing embryo, further improved pregnancy rates of IVF [34]. Nevertheless, manipulations in embryo culture environment may not be benign as it was shown that adverse culture conditions may affect gene expression, imprinting and possibly intrauterine growth resulting in reduced birthweight in IVF newborns [35–37]. These factors should be considered in relation to long-term viability of IVF newborns and the Barker hypothesis. Barker [38] discovered that low-birth weight is related to increased risk of cardiovascular disease and diabetes in adulthood. Culture conditions could therefore have long-term effects on fetal growth, birth weight and even adult health.

Proteomic analysis of the embryonic secretome (proteins produced by the embryo and secreted into the surrounding medium) followed by the identification of specific proteins critical for implantation, enables non-invasive assays to unravel the complexity of embryo physiology, optimize embryo culture media and select the optimal embryo for transfer [37]. This information may enhance the capacity to increase IVF success and prevent genetic disease in a non-invasive fashion.

Embryo Biopsy and PGD/PGS

Embryo biopsy (also known as blastomere biopsy) is a technique that is performed by removal of one or two cells (blastomeres) from the 6 to 8 cell pre-embryo stage for the purpose of preimplantation analysis. On the third day following fertilization, the embryo is at the cleavage stage, and a cell may be carefully removed for genetic analysis. With the embryo maintained in position by gentle suction of the holding pipette, an opening in the outer shell called the zona pellicuda is made using a micro needle. Afterward a new micropipette is used to remove a cell by means of aspiration. At this early stage of embryo development, all of the cells have the same potential for development, therefore, removal of a cell from the embryo is not detrimental and the embryo should continue to develop following the procedure. The cells that have been removed are then tested for a particular genetic disorder.

The possibility of embryo sexing was widely applied in veterinary medicine before being in demand by human [39]. Preimplantation genetic diagnosis (PGD) is a method by which embryos formed through IVF can be tested for single-gene disorders or chromosome abnormalities prior to embryo transfer. This enables couples to significantly improve their chances of having a healthy child. PGD is a complex combination of various technologies that requires close collaboration of a team of specialists for optimal patient care and is now an important addition to conventional prenatal diagnosis for genetic disorders. Until now, it was thought reasonable to limit the use of preimplantation diagnosis to disorders for which prenatal diagnosis was already generally accepted. Future use is likely to expand into areas such as the exclusion of embryos with genes that predispose to adult onset disorders, for which requests for prenatal diagnosis are more unusual. Some people with a genetic predisposition for certain cancers may choose to have preimplantation diagnosis despite the variable penetrance, lethality, and age of onset, and this may even become the largest indication for referral [40]. The range of diseases and susceptibilities for which PGD may be performed continually increases in response to requests from patients and genetic advances provided that genealogic information is available.

Screening for preconception gender selection by using highly reliable PGD and requested by some couples presents ethical issues which need to be addressed. Thus society, in general and particularly the medical community will be forced to take a stand whether this method of sexing should be permitted for social purposes [41].

Preimplantation genetic screening (PGS) for aneuploidy, or simply aneuploidy screening, uses the same approach as PGD to identify embryos with abnormal numbers of chromosomes. Many of these embryos are not viable and will either fail to implant, cause miscarriage or rarely result in fetal or congenital abnormalities. Unfortunately, as aneuploid embryos initially develop normally and cannot be distinguished on the basis of their appearance, they are equally likely to be selected for transfer following IVF. Though it was expected that identifying aneuploid embryos by PGS and transferring only embryos with the normal number of the chromosomes analyzed would improve pregnancy rates, reduce miscarriage rates and reduce the chance of congenital abnormalities associated with aneuploidy, has not been proven. A recent multicenter, randomized, double-blind, controlled trial comparing three cycles of IVF with and without preimplantation genetic screening did not increase but instead significantly reduced the rates of ongoing pregnancies and live births after IVF in women of advanced maternal age [42].

Blastocyst Culture

Blastocyst culture and transfer is an important technique developed for IVF that is receiving renewed attention in order to maximize pregnancy rates while minimizing the risk of multiple pregnancies. With regular IVF, embryos are transferred into the woman's uterus two to three days after fertilization (at the cleavage stage). An alternative technique delays transfer until five to six days after fertilization (at blastocyst stage). This may be better timing and allow the choice of more viable embryos. A review of trials found evidence that more women will have a pregnancy and baby with blastocyst transfer than with regular IVF [43]. There is however, a higher risk that patients would have fewer embryos to freeze and no embryos available for transfer. Another disadvantage of Blastocyst culture is the poor success rates associated with blastocyst cryopreservation. It has taken many years of research to finally develop the right culture medium for blastocyst development and it now appears that blastocyst culture is to be favored in patients with high numbers of eight-cell embryos on Day 3, for whom there is less risk of cycle cancellation.

Single Embryo Transfer

Women undergoing treatment with IVF face a 20-fold increased risk of twins and a 400-fold increased risk of higher order pregnancies (triplets or more) [44]. In 2002, 36% of IVF pregnancies in the USA were multiple fetus pregnancies (SART, 2002). In Europe, where transfer of two embryos are the norm in many centers, 24% of all IVF infants born in 2001 were twins [10]. Multiple pregnancy, including twins, is not a desirable outcome of IVF treatment as it carries an increased risk of very serious consequences including premature birth and cerebral palsy.

Having only a single embryo transferred could prevent these complications. Despite awareness of the risks of twins, patients as well as practitioners remain uncertain about adopting a single embryo transfer policy in routine clinical practice and continue to question its clinical and cost effectiveness [45]. Currently, single embryo transfers are rare, accounting for only 6.2% of all fresh transfers in the USA [46] and 12% of fresh transfers in Europe. Within Europe, there are enormous variations in practice, with rates varying from 30.5% in Finland to 7.3% in the UK [10].

Ideally, it would be useful to be able to identify the embryo with the highest chance of implantation. Techniques such as metabolomics that can assess embryo quality in culture medium may ultimately lead to non-invasive tests for oocyte or embryo quality providing information as to embryo developmental competence. Once fully validated, this new approach is expected to revolutionize embryo selection, leading to improved implantation rates and higher probabilities of success using elective single embryo transfer [47, 48].

Ultrasound Guided Embryo Transfer

It appears to be that the rate-limiting step in IVF success now lies after embryo culture. Indeed most couples can expect to obtain successful fertilization and embryo development. Nevertheless only one in four embryos will implant [48, 49]. One area that is recognized to be very important for IVF success rates is the embryo transfer technique. The use of ultrasound guidance for proper catheter placement in the endometrial cavity has been suggested as a means of significantly improving the technique of embryo transfer. The technique previously used was to insert the embryo transfer catheter into the uterine cavity until reaching the fundus, the top portion opposite the cervix. This technique is now improved by the possibility of examining the endometrial cavity along with the transfer catheter in order to perform the embryo transfer without touching the fundus. Moreover, it was shown that localizing the transfer further away from the fundus increases pregnancy rates. Studies now show a clear advantage from ultrasound-guided embryo transfer which significantly increases the chance of embryo implantation and clinical pregnancy [48–50].

Endometrial Biopsy to Test Endometrial Receptivity

A serious factor in the success of assisted reproduction is endometrial receptivity. Though high quality embryos can be obtained in many patients, two out of three such embryos will not implant in utero. It is surprising that the human endometrium is so hostile to embryo implantation during most of the month. The exception lasts about two days and is termed the 'window of implantation'. Endometrial biopsy, performed during a mock cycle can help determine endometrial receptivity by putative markers of implantation in the endometrium including the presence of pinopods and adhesion molecules as well as other factors [51]. Various factors added to the media are being tested to evaluate whether they could improve embryo implantation rates [52, 53]. Surprisingly, it appears that such a biopsy from the endometrium may increase implantation rates [54, 55].

Oocytes Cryopreservation

Aggressive approaches to cancer diagnosis and treatment now enable high survival rates in many oncological and hematological diseases. The most significant side effect of this treatment is the irreversible damage to the gonads. Whereas young men are routinely referred to sperm freezing prior to chemotherapy, young women's options are more difficult. The best option in terms of success rates requires hormonal stimulation for oocytes aspiration. Retrieved oocytes are preferably fertilized and the resulting embryos are slow frozen on day 2–3 of development. Unfortunately for single women unfertilized oocytes are more sensitive to damage by the freezing process. This is mostly due to the damage by ice on intracellular organelles. Cryoprotectants are utilized to enable oocytes to be frozen albeit at low success rates. New cryopreservation processes such as vitrification may significantly increase the survival, fertilization and pregnancy rates obtained from cryopreserved-thawed oocytes [56]. We are optimistic that in the near future the success rate with oocytes cryopreservation will increase and enable this technology to replace the need to cryopreserve human embryos.

As occurs commonly in the field of assisted reproduction, every new technology opens new applications. Some are more controversial. Such is the case for the request by women to freeze some of their oocytes for later use. The reason for this request is due to the physiological phenomenon of ovarian ageing rendering women at severely reduced fertility potential when they reach their early forties [57–59]. The tendency to delay parenthood in modern society thus imposes a great stress on women who are still childless in their late thirties and now have to consider their options. From this point of view it would make sense for these women to cryopreserve their oocytes at a young age in order to thaw them at the time they are ready to start a family. The cryopreservation of oocytes for social purposes raises serious ethical and practical questions. Nevertheless, this issue needs to be addressed by society in order that guidelines could be composed and followed.

Ovarian Cryopreservation and Transplantation

Another option, which is currently offered, is the possibility of cryopreserving slivers of ovarian cortex. The advantage of this approach is that it does not require any hormonal induction of ovulation and thus it can be performed at short notice.

Indications for ovarian cryopreservation include young female patients prior to gonadotoxic therapy when there is no time or possibility for transvaginal sonographic aspiration of oocytes or for prolonged induction of ovulation. Though a few pregnancies have been reported, this method is still considered to be experimental. Obstacles to success appear to be the ischemia to the grafted ovarian cortex. Research on whole ovary transplantation may enable us to resolve this problem [60–62].

Gestational Carrier Pregnancy

Gestational Carrier Pregnancy is one of the most complicated and expensive processes in ART. Oocytes, harvested from the wife, are fertilized with the husband's sperm in the laboratory and the resulting embryo is then transferred into the gestational carrier's uterus at a specific time during the cycle. Most centers give the gestational carrier hormones to help prime and time the uterus for conception. Indications for a gestational carrier which prevent the wife from carrying the pregnancy include previous hysterectomy, severe fibroids, uterine congenital abnormalities or damage from infection, life threatening medical conditions such as cardiac disease, renal disease, diabetes, active systemic lupus erythematosus (SLE) or estrogen dependent cancers. Other possible indications include numerous failed IVF attempts, emotional factors or physical disabilities. The main difficulty with Gestational Carrier Pregnancy has to do with the need of the carrier of the pregnancy to part with the newborn at delivery. Physicians should carefully examine all relevant issues related to surrogacy, including medical, ethical, legal, and psychologic aspects [63].

Use of Human Embryos for Research – Stem Cells

The potentials of therapeutic cloning could enable treatments for diabetes or chronic cardiac and neurological diseases. The use of supernumerary healthy human embryos or abnormal PGD embryos for embryonic stem cell research has been permitted in Israel. Strict international guidelines are required in order to prevent the creation of embryos for research or the risk that patients would be hyperstimulated so that embryos for research would be available.

Posthumous Conception

Conception and death attract interest and curiosity because they lead to a blurring of the very boundaries between life and death. The issues raised by posthumous assisted reproduction are the most challenging, difficult and sensitive that are likely to be encountered in the field of medicine, let alone reproductive medicine. Taking and using gametes and embryos in posthumous assisted reproduction is almost inevitably emotionally charged, and entails complex moral, ethical and legal concerns [64].

Women in Israel can request to harvest the sperm of their dead husbands, even if the men had not given their prior consent. Those guidelines apply to life-partners, as well as wives, but not to parents or other relatives. Sperm must be removed from a man's body within 24 hours, if it is to be frozen and used for impregnation and thus clinics may remove sperm immediately upon request. The widow will then have to submit an official legal request to the courts before she can use the sperm to become pregnant.

Conclusion

Whereas medicine deals with diagnosis and clinical or surgical treatment of disease, the field of reproductive medicine finds itself always being checked for new boundaries at the frontier of science and ethics. Assisted reproductive technologies are only 30 years old and the majority of work still deals with helping married couples, consisting of a healthy woman and a healthy man, fulfill their dream of having a child. However, technology enables the traditional family roles to be re-evaluated by separating the source of the female and male gametes from the gestational carrier. This results in many variable ways to achieve pregnancy. Moreover, moving the first few days of embryonic development to the laboratory has enabled the examination of the genetic health of the pre-implantation embryo. The risks of selecting out not only for disease but to choose gender or specific genetic traits are already with us. It is therefore crucial that the public in general and policy makers in particular be familiar with the advancements in science and medicine pertaining to the fascinating field I am privileged to practice, the field of reproductive medicine.

Acknowledgments I would like to thank Mrs. June Scher for English editing.

References

1. Steptoe PC, Edwards RG. Birth after the reimplantation of a human embryo. *Lancet* 1978;2(8085):366.
2. Edwards RG, Steptoe PC, Purdy JM. Fertilization and cleavage in vitro of preovulator human oocytes. *Nature* 1970;227(5265):1307–9.
3. Steptoe PC, Edwards RG. Reimplantation of a human embryo with subsequent tubal pregnancy. *Lancet* 1976;1(7965):880–2.
4. Edwards RG, Sharpe DJ. Social values and research in human embryology. *Nature* 1971;231(5298):87–91.
5. Kirby MD. Bioethics of IVF – the state of the debate. *J Med Ethics* 1984;10(1):45–8.
6. Cohen J, Fehilly CB, Fishel SB, Edwards RG, Hewitt J, Rowland GF, et al. Male infertility successfully treated by in-vitro fertilisation. *Lancet* 1984;1(8388):1239–40.
7. Greenfeld DA. Gay male couples and assisted reproduction: should we assist? *Fertil Steril* 2007;88(1):18–20.
8. Boodman G. Fatherhood by a new formula: using an egg donor and a gestational surrogate, some gay men are becoming dads-and charting new legal and ethical territory. *Washington Post* 2005.

9. Woodward BJ, Norton WJ. Lesbian intra-partner oocyte donation: a possible shake-up in the Garden of Eden? *Hum Fertil (Camb)* 2006;9(4):217–22.
10. Andersen AN, Goossens V, Ferraretti AP, Bhattacharya S, Felberbaum R, de Mouzon J, et al. Assisted reproductive technology in Europe, 2004: results generated from European registers by ESHRE. *Hum Reprod* 2008; 23(4): 756–71.
11. Wright VC, Chang J, Jeng G, Chen M, Macaluso M. Assisted reproductive technology surveillance – United States, 2004. *MMWR Surveill Summ* 2007;56(6):1–22.
12. Baird DT, Collins J, Egozcue J, Evers LH, Gianaroli L, Leridon H, et al. Fertility and ageing. *Hum Reprod Update* 2005;11(3):261–76.
13. Greb RR, Behre HM, Simoni M. Pharmacogenetics in ovarian stimulation – current concepts and future options. *Reprod Biomed Online* 2005;11(5):589–600.
14. Fauser BC, Diedrich K, Devroey P. Predictors of ovarian response: progress towards individualized treatment in ovulation induction and ovarian stimulation. *Hum Reprod Update* 2008;14(1):1–14.
15. Lunenfeld B, Rabau E, Sulomivici S, Eshkol. [Treatment of amenorrhea by gonadotropic substances from women's urine.]. *Harefuah* 1963;64:289–92.
16. Van Wely M, Westergaard LG, Bossuyt PM, Van der Veen F. Human menopausal gonadotropin versus recombinant follicle stimulation hormone for ovarian stimulation in assisted reproductive cycles. *Cochrane Database Syst Rev* 2003(1):CD003973.
17. Palermo G, Joris H, Devroey P, Van Steirteghem AC. Pregnancies after intracytoplasmic injection of single spermatozoon into an oocyte. *Lancet* 1992;340(8810):17–8.
18. Jain T, Gupta RS. Trends in the use of intracytoplasmic sperm injection in the United States. *N Engl J Med* 2007;357(3):251–7.
19. Bhattacharya S, Hamilton MP, Shaaban M, Khalaf Y, Seddler M, Ghobara T, et al. Conventional in-vitro fertilisation versus intracytoplasmic sperm injection for the treatment of non-male-factor infertility: a randomised controlled trial. *Lancet* 2001;357(9274):2075–9.
20. Leunens L, Celestin-Westreich S, Bonduelle M, Liebaers I, Ponjaert-Kristoffersen I. Follow-up of cognitive and motor development of 10-year-old singleton children born after ICSI compared with spontaneously conceived children. *Hum Reprod* 2008;23(1):105–11.
21. Kurinczuk JJ. Safety issues in assisted reproduction technology. From theory to reality – just what are the data telling us about ICSI offspring health and future fertility and should we be concerned? *Hum Reprod* 2003;18(5):925–31.
22. Fasouliotis SJ, Safran A, Porat-Katz A, Simon A, Laufer N, Lewin A. A high predictive value of the first testicular fine needle aspiration in patients with non-obstructive azoospermia for sperm recovery at the subsequent attempt. *Hum Reprod* 2002;17(1):139–42.
23. Lewin A, Reubinoff B, Porat-Katz A, Weiss D, Eisenberg V, Arbel R, et al. Testicular fine needle aspiration: the alternative method for sperm retrieval in non-obstructive azoospermia. *Hum Reprod* 1999;14(7):1785–90.
24. 2006 Guidelines for Gamete and Embryo Donation. *Fertil Steril* 2006;86(5 Suppl):S38–50.
25. Heng BC. The advent of international 'mail-order' egg donation. *Bjog* 2006;113(11):1225–7.
26. Heng BC. Factors influencing the reimbursement rate of egg donation within a competitive free-market system. *Reprod Biomed Online* 2007;15(1):16–8.
27. Cohen J. Procreative tourism and reproductive freedom. *Reprod Biomed Online* 2006;13(1):145–6.
28. EU. DIRECTIVE 2004/23/EC OF THE EUROPEAN PARLIAMENT AND OF THE COUNCIL on setting standards of quality and safety for the donation, procurement, testing, processing,preservation, storage and distribution of human tissues and cells. In: PARLIAMENT E, editor: EUROPEAN PARLIAMENT, 2004.
29. Craft I, Flyckt S, Heeley G, Layland S, Thornhill A, Kelada E. Will removal of anonymity influence the recruitment of egg donors? A survey of past donors and recipients. *Reprod Biomed Online* 2005;10(3):325–9.
30. Landau R, Weissenberg R, Madgar I. A child of "hers": older single mothers and their children conceived through IVF with both egg and sperm donation. *Fertil Steril* 2008; 90(3): 576–83.

31. Buckett WM, Chian RC, Holzer H, Dean N, Usher R, Tan SL. Obstetric outcomes and congenital abnormalities after in vitro maturation, in vitro fertilization, and intracytoplasmic sperm injection. *Obstet Gynecol* 2007;110(4):885–91.
32. Cha KY, Chian RC. Maturation in vitro of immature human oocytes for clinical use. *Hum Reprod Update* 1998;4(2):103–20.
33. Alexandre H. A history of mammalian embryological research. *Int J Dev Biol* 2001;45(3):457–67.
34. Gardner DK, Lane M. Culture of viable human blastocysts in defined sequential serum-free media. *Hum Reprod* 1998;13 Suppl 3:148–59;discussion 160.
35. Schieve LA, Meikle SF, Ferre C, Peterson HB, Jeng G, Wilcox LS. Low and very low birth weight in infants conceived with use of assisted reproductive technology. *N Engl J Med* 2002;346(10):731–7.
36. Behr B, Wang H. Effects of culture conditions on IVF outcome. *Eur J Obstet Gynecol Reprod Biol* 2004;115 Suppl 1:S72–6.
37. Katz-Jaffe MG, Gardner DK. Embryology in the era of proteomics. *Theriogenology* 2007;68 Suppl 1:S125–30.
38. Barker DJ. Fetal origins of coronary heart disease. *BMJ* 1995;311(6998):171–4.
39. Beardsley T. Embryo sexing. Cattle now, people next? *Nature* 1983;304(5924):301.
40. Braude P. Preimplantation diagnosis for genetic susceptibility. *N Engl J Med* 2006;355(6):541–3.
41. Schenker JG. Gender selection: cultural and religious perspectives. *J Assist Reprod Genet* 2002;19(9):400–10.
42. Mastenbroek S, Twisk M, van Echten-Arends J, Sikkema-Raddatz B, Korevaar JC, Verhoeve HR, et al. In vitro fertilization with preimplantation genetic screening. *N Engl J Med* 2007;357(1):9–17.
43. Blake DA, Farquhar CM, Johnson N, Proctor M. Cleavage stage versus blastocyst stage embryo transfer in assisted conception. *Cochrane Database Syst Rev* 2007(4):CD002118.
44. Martin PM, Welch HG. Probabilities for singleton and multiple pregnancies after in vitro fertilization. *Fertil Steril* 1998;70(3):478–81.
45. Pandian Z, Templeton A, Serour G, Bhattacharya S. Number of embryos for transfer after IVF and ICSI: a Cochrane review. *Hum Reprod* 2005;20(10):2681–7.
46. SART. Assisted Reproductive Technology Success Rates for 2002. In: Prevention CfDCa, editor. *National Summary and Fertility Clinic Reports*. Washington,DC: US Department of Health and Human Services, 2004.
47. Patrizio P, Fragouli E, Bianchi V, Borini A, Wells D. Molecular methods for selection of the ideal oocyte. *Reprod Biomed Online* 2007;15(3):346–53.
48. Brown JA, Buckingham K, Abou-Setta A, Buckett W. Ultrasound versus 'clinical touch' for catheter guidance during embryo transfer in women. *Cochrane Database Syst Rev* 2007(1):CD006107.
49. Buckett WM. A meta-analysis of ultrasound-guided versus clinical touch embryo transfer. *Fertil Steril* 2003;80(4):1037–41.
50. Abou-Setta AM, Mansour RT, Al-Inany HG, Aboulghar MM, Aboulghar MA, Serour GI. Among women undergoing embryo transfer, is the probability of pregnancy and live birth improved with ultrasound guidance over clinical touch alone? A systemic review and meta-analysis of prospective randomized trials. *Fertil Steril* 2007;88(2):333–41.
51. Achache H, Revel A. Endometrial receptivity markers, the journey to successful embryo implantation. *Hum Reprod Update* 2006;12(6):731–46.
52. Revel A, Helman A, Koler M, Shushan A, Goldshmidt O, Zcharia E, et al. Heparanase improves mouse embryo implantation. *Fertil Steril* 2005;83(3):580–6.
53. Simon A, Safran A, Revel A, Aizenman E, Reubinoff B, Porat-Katz A, et al. Hyaluronic acid can successfully replace albumin as the sole macromolecule in a human embryo transfer medium. *Fertil Steril* 2003;79(6):1434–8.

54. Barash A, Dekel N, Fieldust S, Segal I, Schechtman E, Granot I. Local injury to the endometrium doubles the incidence of successful pregnancies in patients undergoing in vitro fertilization. *Fertil Steril* 2003;79(6):1317–22.
55. Zhou L, Li R, Wang R, Huang HX, Zhong K. Local injury to the endometrium in controlled ovarian hyperstimulation cycles improves implantation rates. *Fertil Steril* 2007.
56. Oktay K, Cil AP, Bang H. Efficiency of oocyte cryopreservation: a meta-analysis. *Fertil Steril* 2006;86(1):70–80.
57. Gougeon A. Dynamics of follicular growth in the human: a model from preliminary results. *Hum Reprod* 1986;1(2):81–7.
58. Gougeon A. Regulation of ovarian follicular development in primates: facts and hypotheses. *Endocr Rev* 1996;17(2):121–55.
59. Gougeon A, Ecochard R, Thalabard JC. Age-related changes of the population of human ovarian follicles: increase in the disappearance rate of non-growing and early-growing follicles in aging women. *Biol Reprod* 1994;50(3):653–63.
60. Revel A, Schenker J. Ovarian tissue banking for cancer patients: is ovarian cortex cryopreservation presently justified? *Hum Reprod* 2004;19(1):14–9.
61. Weintraub M, Gross E, Kadari A, Ravitsky V, Safran A, Laufer N, et al. Should ovarian cryopreservation be offered to girls with cancer. *Pediatr Blood Cancer* 2007;48(1):4–9.
62. Revel A, Elami A, Bor A, Yavin S, Natan Y, Arav A. Whole sheep ovary cryopreservation and transplantation. *Fertil Steril* 2004;82(6):1714–5.
63. ACOG committee opinion number 397, February 2008: surrogate motherhood. *Obstet Gynecol* 2008;111(2 Pt 1):465–70.
64. Bahadur G. Death and conception. *Hum Reprod* 2002;17(10):2769–75.

Chapter 2
The Medicalization of Reproduction – A 30 Year Retrospective

Søren Holm

There is a line of argument which attempts to support a general prohibition of pre-implantation diagnosis. [. . .] The argument seeks to establish that pre-implantation diagnosis is harmful because it further medicalizes and technologizes human reproduction. According to this way of looking at reproductive techniques, they are harmful because they transform something which has (appropriately) been seen as natural into something where technology, medical knowledge, and human choice become essential. It seems relevant to ask 'but what is wrong with technologizing reproduction? [1, p.188]

Introduction

Now, 30 years after the birth of Louise Brown, the first IVF baby, it is relevant to assess how the Assisted Reproductive Technologies (ART) have affected reproductive practices and our view of reproduction. Has ART medicalized and technologised reproduction, and if so does that raise any ethical problems? Are there, for instance harmful effects of medicalization, or has the developments reduced the freedom of some people?

In the quote above from 1998 I focused on the possible technologising effects of ART, but here I want to focus on the possible medicalising effects.

Since the invention of In Vitro Fertilisation (IVF) there has been an explosion in different kinds of ART each presumably responding to the reproductive needs and desires of a particular subset of would be reproducers. Intracytoplasmatic Sperm Injection allows men with very few, or only immature sperm to reproduce; the use of surrogate gestational mothers responds to the need of women who cannot gestate or of male homosexual couples; Preimplantation Genetic Diagnosis (PGD) for genetic disorders helps families with known genetic problems to avoid prenatal diagnosis and abortion etc. etc. If we further add the possibilities of using donor sperm and

S. Holm (✉)
Cardiff Law School, ESRC Centre for the Economic and Social Aspects of Genomics (Cesagen) & Section for Medical Ethics, University of Oslo, Oslo, Norway
e-mail: holms@cardiff.ac.uk

F. Simonstein (ed.), *Reprogen-ethics and the Future of Gender*, International Library of Ethics, Law, and the New Medicine 43, DOI 10.1007/978-90-481-2475-6_3,

eggs, of freezing embryos and eggs, and possibly in the future of creating gametes from stem cells or performing reproductive cloning or germ line genetic modification we have a situation where there is a possible ART solution to the reproductive needs and desires of almost any combination of would be reproducers. The extent of these present and future technical possibilities is laid out in detail in Chapter 1 of this book.

It may cost you (a lot!), but if you want to reproduce or reproduce in a specific way there will be an ART available enabling you to do so. The development of all of these current (and future) ART has thus expanded the scope of reproductive liberty enormously, and that must surely be a good thing!? If for no other reason simply because every expansion of an important liberty is a *prima facie* good thing.

In the following I want to problematise this view partly through a discussion of what the reproductive liberties in question actually amount to in terms of effective reproductive freedom, and partly through a discussion of whether or not ART leads to a problematic medicalization of reproduction. I will not provide an analysis of the scope and content of the putative right to procreate, partly because an excellent analysis is provided in Chapter 3 by Marleen Eijkholt, partly because the analysis is only tangential to the question of medicalization.

But before moving to the analysis of medicalization it is important to state two caveats about the analysis up front. Both of these are related to the, in my view obviously true observation that a discussion of reproductive liberties and medical and other practices in relation to ART does not centrally engage what are quantitatively and morally the most important questions in any discussion of reproductive rights – the question of how the reproductive rights of women in resource poor, often strongly patriarchal societies can be secured. In that context what is often needed is access to long extant technologies making it possible for women to choose not to reproduce, and increasing, not decreasing the medicalization of reproduction. The United Nations' Convention on the Elimination of All Forms of Discrimination Against Women does give women extensive reproductive rights in Article 16, but these rights are in practice not secured in many countries in the world.

Liberty, Effective Freedom and the Social Embedding of ART

It is, as stated above obvious that the development in ART has produced an enormous increase in reproductive options and therefore also in those societies that allow a wide range of ART an increase in the negative liberties of people (here using negative liberty in the sense defined by Berlin as freedom from external constraint [2]). This increase has come about primarily because ART makes new kinds of reproductive projects possible (i.e. removes biological obstacles to reproduction) and secondarily because the development of ART and their successful application has led to, or at least contributed to changes in societal views and regulations (e.g. in relation to homosexual parenting). The increase in negative liberty does in principle accrue to everyone, both those who want to reproduce and who take advantage of the liberties afforded them, and those who merely hold the new liberties as unused and perhaps undesired options. Both the involuntarily infertile and the voluntarily celibate have had their liberties increased by ART, although only one of these groups is likely to exploit those liberties.

But having a negative liberty is not the same as having effective freedom [3]. As already Anatole France remarked both the rich and the poor man is free (i.e. has negative liberty) to sleep under the bridges of Paris. But whereas they both have this negative liberty and the negative liberty to stay in the Ritz Hotel, only the rich man has effective freedom to do the latter because only the rich man possesses the resources to act effectively and exploit his liberty.

A full exploitation of the new reproductive opportunities is similarly only available to those who are in a position to pay privately. No health care system funds unlimited access to all those ART that are legally available within that particular jurisdiction, and very few fund any access to those ART that are only legally available outside the jurisdiction (Norway until recently being an interesting exception in relation to PGD. PGD was illegal in Norway but never the less funded by the Norwegian state for some couples at institutions outside of Norway).

Effective use of ART is also limited by government regulations and by the medical profession (see below in the section on medicalization) for those who do not have the financial and other (e.g. informational) resources to seek out "mavericks" or have treatment outside of their own country.

For those seeking to reproduce it can therefore be very difficult to convert their ostensible reproductive liberties into effective reproductive freedom; and even if they can get access to ART the relatively low success rates for many techniques entail that the effective freedom that they have attained is in essence only a freedom to attempt reproduction, not a freedom to reproduce.

In this context it is important to remember that regulation of ART is not undesired by those who make use of the techniques. Many of them want to achieve goals that are not purely biological, but also legal and social. Some do, for instance not only want to have a child, they also want the state to declare that the child is legally their child and they are its parents, even though they are not biologically related to the child. And most want the state to regulate ART providers in some ways with regard to competence and safety, but competence and safety cannot be regulated without sometimes circumscribing the liberties of both providers and users.

More paradoxically the increase in procreative liberties created by the developments of ART has indirectly decreased the effective freedom of some people. This situation has come about because the broad social acceptance of ART in many countries has changed social default expectations with regard to reproductive choices. Anyone who admits to being infertile and still having some desire for children is likely to be met with the expectation that they should engage in ART treatment. This will lead a minority to pursue such treatments against their own better judgement, and may negatively affect the wellbeing of a much larger group who will feel under pressure to defend their decision not to use ART.

In an article on what he calls the "argument from coercion" Petersen discusses the argument that women are socially coerced to use ART and that this is a good reason to ban ART [4]. He rightly rejects this argument, primarily because there is no good reason to believe that the choices in question are not autonomous. As part of this rejection he states an interim conclusion that he, unfortunately does not develop any further. This conclusion is that "First, if women are being coerced to desire to use AR, we should eradicate the coercive elements in pro-natalist ideology, not access to ART" [4, p. 81]. This conclusion is a version of a fairly common argumentative move in bioethics stating that if a certain negative effect of an act or policy can be traced to some kind of prejudice or ideology in society, then that is not a reason not to perform the act or introduce the policy, but a reason to eradicate the prejudice or ideology. Just as in Petersen's article it is, however seldom explicated how such an eradication should be achieved, or who might be responsible for achieving it. So, let us briefly depart from the norm in bioethics and ask ourselves how the negative effects of ART on some people's effective freedom and welfare can be ameliorated or eradicated, and who might be responsible in this context.

The negative effects we have identified above are not due to ART *per se*, but to ART becoming the *de facto* social default in certain situations. This helps us to identify the first actions to take and the first responsible agents. It will be difficult to target every single individual who engages in what we could call "stereotypical fertility talk" perpetuating the problematic social conceptions, but we can target those who engage in such talk because it actively promotes their own interests. We can essentially ask the classical legal question "*Qui Bono*?" Pre-eminent in the class of agents that benefit from presenting use of ART as the default option are fertility clinics and fertility specialists. So maybe we should begin the process of increasing the effective freedom and wellbeing of those who do not want to use ART by requiring clinics and professionals to be scrupulously honest in what they explicitly and implicitly promise and not to emphasise their particular solution to reproductive issues as better or more appropriate than other solutions.

Next we might target magazines that run tear jerking infertility human interest stories, or we might take a closer look at bioethicists and the extent to which their writings perpetuate the idea that childlessness is in general a great misfortune that should be solved by ART.

In the scrutiny of the communications of these groups it will not be sufficient to look at whether the communications are true; we will also have to look at whether they employ a misleading rhetoric. It is always important to remember that it is possible to mislead by telling the truth.

ART and the Medicalization of Reproduction

The concept of medicalization has several meanings and occurs within different theoretical discourses. It can describe a process, as in "reproduction is becoming increasingly medicalized" or a state of "medicalization", and if it is given a theoretical backing it is often within either a Foucauldian or Habermasian frame of reference. Originally, however the concept of medicalization was first put forward in American sociology on the basis of Parsons' analysis of medicine as an institution of social control (e.g. in his analysis of the patient role) [5] or with a foundation in labelling theory [6]. Another possible meaning of medicalization is in discussions of inter-professional relations and boundary drawing where the term is used in a negatively value-laden way to describe the (illegitimate) usurpation by the medical profession of areas of activity that are properly the concern of other professions.

In this paper the concept will be used in an as a-theoretical form as possible to refer to the process that occurs when a given area of human activity is defined socially as falling under the proper purview of medicine, and to any state that such a process might lead to. For the theoretically minded this is obviously more closely connected to Habermas' ideas about the encroachment of the systemworld upon the lifeworld and the subsequent colonisation of the lifeworld [7] than it is to Foucault. The concept of medicalization as used here does, however not have the inherent value-ladenness of Habermas' colonisation concept. That a certain area of human activity becomes medicalized may be positive, neutral or negative from an ethical point of view (see more below).

The precise features of medicalization are disputed and may well differ from area to area, but at its core medicalization involves allowing the medical profession sole or pre-eminent power to define a given issue, decide whether or not it is a problem, and define the appropriate (therapeutic) solutions to the problem. We can thus distinguish between three levels or kinds of medicalization: conceptual (a medical frame of reference is being used), institutional (the issues are dealt with in medical institutions), and interactional (the medical profession is directly involved in dealing with the issue) [8]. Because medicalization gives the profession definatory power it will often in practice also lead to the appropriate solutions being defined within a medical frame, and such solutions will often involve the necessary involvement of medical practitioners. In short the process of medicalization allows the medical profession to take control of an area of life that other parties, often ordinary people previously controlled. In some cases this is clearly a good thing. The medicalization of severe stomach pains has undoubtedly produced immense benefit over time, as has the medicalization of fevers. In other cases the benefits are more questionable (e.g. the medicalization of sexuality) or there are perhaps overall negative effects (e.g. in the case of the medicalization of shyness as "social anxiety disorder" [9]).

It is not a necessary precondition of medicalization that the medical profession has effective therapeutic solutions in a given area, the area of mental disturbance was for instance medicalized long before effective treatments became available, but having effective options to offer is clearly conducive to medicalization. It is also not necessary that the area in question is viewed as socially problematic deviant at the beginning of the process, although it may help the process get started [9–11].

Has reproduction become more medicalized because of ART and is that an unalloyed good?

There is no doubt that reproduction was already significantly medicalized in the western world before the advent of IVF, but what was medicalized was primarily contraception, abortion and the middle and later stages of pregnancy and childbirth [12]. The process of conception was mainly outside the scope of medical interest, partly because very few effective treatments where available, artificial insemination by donor being an exception.

With the invention of IVF and further ART the technical powers of the medical profession rapidly increased and the process of human conception, or perhaps more accurately lack thereof became rapidly medicalized. It is now estimated that as many as 4.2% of all births in Denmark were the result of some use of ART in 2004. This was the highest proportion in Europe in 2004, and is likely to have gone up since [13].

But the most important development from a perspective of medicalization was that the new techniques also allowed the medical profession to offer services and thereby medicalize areas of reproduction where no clinical problem had previously been defined, because the problem in question had either been seen as unsolvable (e.g. facilitating the reproduction of men with azoospermia), or as a social problem (e.g. facilitating the reproduction of lesbian couples or single women).

There is thus little doubt that the development of IVF and subsequent ART has contributed significantly to the medicalization of reproduction, as has also been pointed out by numerous other writers [14]. Is this medicalization ethically problematic?

To answer that question we need to look more closely at what the effects of the medicalization has been. It has, as discussed above increased reproductive liberty and the effective reproductive freedom of some, but it has done so in very specific ways. Many of the new liberties cannot be utilised directly by those seeking to reproduce but must be accessed through a medical practitioner who acts as gatekeeper and is often also involved in the actual ART process. That this should be so is not logically necessary, we could in principle have arranged ART provision differently, but because of the medicalization of the area the medical profession was always likely to be given this role by societal default.

Here we may in passing note that at the moment we involve someone in a causally pivotal role in reaching a particular goal, we must also almost always involve that someone in the moral evaluation of that goal. Human beings are pre-eminently moral actors and cannot be reduced to a-moral agents or automatons in the pursuit of my projects. To the extent that ART technically involve other persons than those wishing to reproduce those other persons will always have to decide for themselves whether this is a reproductive project they should contribute to. This is true whether or not those we involve are medical professionals.

But let us return to the issue of medicalization. The fact that the medical profession is involved has practical implications that again, although not logically necessary (because the existence of a medical profession of the kind that exists in the western world is not logically necessary) are nevertheless practically certain to occur. What are these implications and how can they be evaluated?

The first implication is that medical professionals do not come as stand alone technical experts or moral actors. The medical professional is embedded in the medical profession, and usually also in some sub-specialty of that profession and this means that the medical professional comes with both implicit and explicit values, regulations and guidelines attached.

The most important implicit value is undoubtedly that health is the most important goal that an individual can pursue [15]. Unhealthy reproductive choices will therefore be frowned upon, and may be actively resisted or simply not offered as part of ART. This will apply to both choices that are unhealthy for those reproducing, and choices that are unhealthy for the child to be. This is, for instance evident in the current move towards reducing multiple pregnancies by the transfer of only one embryo. In many cases those reproducing are not given the option of balancing success rates against the risks inherent in multiple pregnancies, that balancing is done by the profession. For those who believe that reproductive liberty should be overriding, this is clearly an ethically problematic aspect of medicalization (see for instance Purdy's similar argument in relation to fetal reduction [16]).

Conversely, defining childlessness as a medical problem and thereby by implication a problem with importance in relation to health makes it more difficult to understand why anyone would decide not to have their problem solved by medical means. We may agree in the abstract that people have a right based in respect for self-determination to decline medical treatment for their medical problems, but the framing of the problem as medical is conducive to thinking that people who decline treatment are more odd than if we had defined the problem as social or existential.

Guidelines and regulations for ART are not in themselves a problem, they are in some sense necessary as mentioned above, but medicalization entails that the medical profession becomes a dominant force in developing the guidelines. This in turn entails that the guidelines are unlikely to be based purely on an evaluation of the best interests of users of ART or of society in general. Some aspects of guidelines are likely to be formulated to primarily benefit the profession and not its clients or customers. They will, for instance often make the profession the gatekeeper to the services in question.

Finally, as already alluded to above the institutional medicalization of ART also legitimises the marketing of the techniques by the more entrepreneurially minded members of the profession. Medicalization legitimises the claim that what they are offering are not merely techniques to give people what they want, it is *bona fide* medical treatment and thereby much more important! The fact that there is no special reason to treat treatment as more important than other ways of solving the same problem, does nothing to detract from the perception that treatment is special [17].

Conclusion

In this paper I have argued that although it is true that the developments of ART over the last 30 years have expanded the scope of reproductive liberty, and that such an expansion is a *prima facie* good thing, the current use of ART is not unproblematic. While ART may have increased abstract liberty they have done much less to increase effective reproductive freedom overall; and the development of ART has decreased the effective freedom of some people.

ART has also contributed significantly to the ongoing medicalization of reproduction. The options that are open to would be reproducers are, except perhaps for the obscenely wealthy, significantly constrained not only by democratically decided laws and regulation but by the medical profession and individual doctors.

There are good reasons to try to combat the negative effects of the social embedding of ART and the medicalization of ART. This is true whatever view one takes with regard to the ethical assessment of ART in general, or specific types of ART.

References

1. Holm S. Ethical Issues in Pre-implantation Diagnosis. In: Harris J, Holm S (eds.). *The Future of Human Reproduction*. Oxford: Oxford University Press, 1998. (pp. 176–190).
2. Berlin I. *Four Essays on Liberty*. Oxford: Oxford University Press, 1969.
3. Swift A. *Political Philosophy: A Beginner's Guide for Students and Politicians*. Cambridge: Polity, 2001.
4. Petersen TS. A woman's choice? – On women, assisted reproduction and coercion. *Ethical Theory and Moral Practice* 2004; 7: 81–90.
5. Parsons T. *The Social System*. New York: Free Press, 1951.
6. Conrad P. Medicalization and social control. *Annual Review of Sociology* 1992; 18: 209–232.
7. Habermas J. *The Structural Transformation of the Public Sphere: An Inquiry into a Category of Bourgeois Society*. Trans. Thomas Burger with Frederick Lawrence. Cambridge, MA: MIT Press, 1991.
8. Conrad P, Schneider J. Looking at levels of medicalization: A comment on Strong's critique of the thesis of medical imperialism. *Social Science and Medicine* 1980; 14: 75–79.
9. Conrad P. *The Medicalization of Society*. Baltimore, MD: Johns Hopkins University Press, 2007.
10. Conrad P, Schneider J. *Deviance and Medicalization: From Badness to Sickness*. St. Louis: Mosby, 1980.
11. Rosecrance J. Compulsive gambling and the medicalization of deviance. *Social Problems* 1985; 32: 275–284.
12. Lee E. *Abortion, Motherhood and Mental Health*. New York: Aldine de Gruyter, 2003.
13. Andersen AN, Goosens V, Ferraretti AP, Bhattacharya S, Felberbaum R, de Mouzon J et al. Assisted reproductive technology in Europe, 2004: Results generated from European registers by ESHRE. *Human Reproduction* 2008; 23: 756–771.
14. Callahan JC (ed.). *Reproduction, Ethics, and the Law: Feminist Perspectives*. Bloomington: Indiana University Press, 1995.
15. Crawford R. Healthism and the medicalization of everyday life. *International Journal of Health Services* 1980; 10: 365–388.
16. Purdy L. Women's reproductive autonomy: medicalization and beyond. *Journal of Medical Ethics* 2006; 32: 287–291.
17. Holm S. Infertility, childlessness, and the need for treatment. In: Evans D (ed.). *Creating the Child*. Dordrecht: Kluwer, 1996 (pp. 65–78).

Chapter 3
Conceiving a New Right to Procreate

Marleen Eijkholt

Introduction

The chapters of Holm and Revel in this book demand the establishment of a clear and uncontroversial form for right to procreate. Their chapters imply the need of a framework for regulation, and indicate the need for a basis to judge the developments of reproductive techniques, especially as a result of the changing positions of men and women in the reproductive process. Holm's chapter sets out that having a liberty to choose does not yet amount to an effective freedom, and he questions the effects of medicalisation. His chapter leads to questions about how to we can reconcile liberty and freedom, and, in the light of the medicalisation, which framework we would best place them in. Whereas Holm avoids talking about rights, his considerations and conclusions demand for a basic starting point, a minimum guarantee. Revel elaborates on the state of the art of new reproductive techniques, and his conclusion touches upon the boundaries of law and ethics. Revel's chapter requires us to think about the availability of these techniques; what the shifting position of men and women would mean for this assessment, and whether there is a right to these techniques. This chapter seeks to create the basic starting point that these chapters need: A notion that could harmonise law and ethics better than the current conception, which is vague and we try to avoid referring to as a result of its complexity. This new notion tries to do that by dissecting the process of reproduction until reaching the most basic and bare standard that could provide protection: the capacity-based standard. In a minimalistic conception, my notion is aimed at preventing randomized availability of reproductive techniques, addresses the ambiguities in regulation as a result of their development, as apprehended by Holm and Revel, and offers a framework for the many issues that the field entails.

In both academic literature and popular debate, the concept of a right to procreate is much discussed and disputed. I argue in this paper that some of the controversy

M. Eijkholt (✉)
Centre for Social Ethics and Policy, Institute for Science, Ethics and Innovation,
University of Manchester, Manchester, England, UK
e-mail: marleen.eijkholt@manchester.ac.uk

F. Simonstein (ed.), *Reprogen-ethics and the Future of Gender*, International Library of Ethics, Law, and the New Medicine 43, DOI 10.1007/978-90-481-2475-6_4,

surrounding claims of a right to procreate derives from the obscurity of the term itself. There is no consensus on just what is meant by those deceptively simple three words "right to procreate". Other kinds of alleged rights such as the right to respect for family life, the right to rear a child or found a family are too often taken to be the same as a right to procreate. Different writers claim different things when using the same words. Yet, when trying to define the common core in these claims, many writers have traditionally found that the right to procreate is about having a child to rear. I argue, in this paper, that this approach is paradoxically too narrow and too broad.

As an alternative, I offer a simple and basic definition of the right to procreate. It is this. Everyone has the right to realize and control his or her capacity to procreate, that is the capacity to engage in the process of procreation, or the process in order to conceiving or begetting offspring. Exercising this right means both a prima facie claim not to have his/her capacity to procreate removed against his/her will and that where a person is affected by some remediable problem in his/her capacity to procreate, the right provides that he/she should not be prevented from having access to treatment to overcome that problem. The emphasis within this notion lies on the capacity, and the possibility to control and realize this capacity, which is required to engage in the process, not to be confused with the realization of the conception or fertilization itself [1]. By focusing on the individual's capacity, the notion dissociates itself from the traditional fixation on realizing the fertilization, and the conception of a child, but is indifferent to the outcome of the process. The right that I seek to establish is separate from, and different to, such rights as the right to privacy or bodily integrity, or the right to rear a child. It is a lowest common denominator. The claim for a right based so firmly on the capacity to procreate does not mean that those other rights may not also be advocated.

I argue that the semantic difficulty in the traditional approach is not simply pedantic, but it means that it is harder to reach any consensus on the substance of the right being claimed. I will even go further than that and argue that it prevents us from defining the right to procreate as a fully fledged moral or legal human right. My revised definition seeks to meet the problems arising under the traditional approach and forms a better basis for legislation; it can deal with the interests of the child, has a well delineated scope, and has a gender neutral starting point.

In order to show why a new conception of the right to procreate is desirable, I will, in section "The Traditional View and its Shortcomings", analyze the traditional notion, then set out the new notion and its justification, in section "The Capacity to Procreate", and, finally, in section "Problems of Implementation and how These can be Solved by the New Approach", I will address some of the problems of the traditional notion and demonstrate how the new conception can deal with these.

The Traditional View and its Shortcomings

The reasons why I believe that the current usages of the term 'the right to procreate' tend to obscure, and undermine the usefulness of this notion's content, will be the subject of this first section. Here I delineate the foundation for my claim that a new definition of this purported right is a desideratum.

The Right to Procreate: Difficult to Describe

Since it is difficult to define what the right to procreate is, this first section is not concerned with giving a definition [2]. It will just highlight the problems that arise when trying to describe the right. In the first instance, it is tempting to describe the right to procreate as being about the right to have children. However, this explanation is too simplistic. It does not take account of the subtle distinctions within the various references to the right. For example, the right to have children is different from the right to create offspring. Whereas the former may refer to an adopted child, the latter will only refer to biological descendants.

Appeals to the right to procreate are made in various situations, and they are not always related to a right to have children. For example, the right to procreate can be evoked when talking about fertility techniques: it may be used to claim access to IVF treatments. Yet, on closer scrutiny this claim seems to be more closely related to a right to health care than a right to have offspring, even though having offspring might be an underlying reason for the claim. Further, we come across allusions to the right in discussions about particular uses of reproductive techniques, like in the cases of sex-selection or pre-implantation genetic diagnosis (PGD). However, when examining these claims more closely, they seem more related to a right to decide what sort of offspring to have, than a right to have offspring *per se*.

Further, there is confusion about which practices would fall under the right and which would not. In some models the right to procreate comes into play only in situations in which people actively desire to beget offspring, not in a situation in which there no such desire. Such a 'right' to procreate would then not extend to the issues of contraception, sterilization or abortion. Yet, in other models the right would cover them. The argument is that the right is anchored in self-determination and since these practices are about self-determination either to or not to procreate, the right to procreate would apply to these practices as well.

Sometimes other terms – instead of the right to procreate – are used in discussions on similar or related matters. These include the right to reproductive autonomy [3], the right to procreative freedom [4] and the right to found a family [5]. The alternative terms are sometimes, but not always, used interchangeably with the right to procreate. This causes confusion not only over how these notions relate to the right, but also over its definition. It is argued here that the expressions are employed to refer to the same issues as the right to procreate. The difference is that they are less neutral and evoke a certain point of view towards the issues at stake – specifically, the terms 'autonomy' and 'freedom' are revealing in this respect. They place primary importance on the choice of the individual, have positive connotations and ignore a broader aspect of procreation whilst these notions imply the conception of a child.

The different references also use different foundations for the right. This creates obscurity in the way in which the right exists and how it should be described. Some accounts hold that the right, as such, exists as a moral right, independently of legislation. To talk about it is an 'expression of [a] moral perspective' [6]. Other accounts see it as a legal right, although no jurisdiction has specifically endorsed a right to procreate in unequivocal terms. The claim is that it can be derived, for example, from the right to privacy and family life (Article 8), and the right to found a family (Article 12) of the European Convention on Human Rights (ECHR). Whether these Articles can offer suitable protection for any kind of right to procreate is doubtful.

The Moral Relevance of the Right to Rear a Child on the Traditional View

Despite obscurity in defining the right to procreate, and what is its basis, several authors have traditionally pointed to an apparent meaning. Although they have not been able to define it, they have argued that the moral relevance of the right lies in its intention to protect the 'rearing of the child' [7]. Bonnie Steinbock, for example, holds that the basic value of procreation lies in having offspring to rear [8]. She argues that having children and raising them is important for a person's identity. Accordingly, she finds that a claim to the right to procreate is diminished when a claimant does not have the ability to bring up a child. When a person lacks this ability, he cannot appreciate the importance of the right [9]. Thus, an individual with a learning disability who cannot take care of a child has a 'very much reduced claim' on the right to procreate [8]. The decisive aspect in Steinbock's vision of any right to procreate is parenting. Onora O'Neill also holds that the right to procreate is about bringing up children. The right would be dependent on whether 'they [the parents] are capable of and committed to being present and active for a child' [10]. O'Neill uses the 'right to parent' as its synonym, and characterizes both rights as 'sometimes confined to rearing without the biological reproduction' [11–15]. She argues that if a child is given away after birth, this would be detrimental to the right to procreate, as it would eliminate the possibility of raising the child [11]. Were a child to be taken away after birth, by for example government institutions, this would be an infringement of the right [8]. Similarly, Elizabeth Scott holds that the 'right to procreate is the right to produce one's own children to rear' [16]. Her justification lies in the intention to rear and this determines the right's fundamental status. If the intention is not there, there is no right to procreate.

Legal sources seem to support the idea that the significance of the right to procreate is linked with the idea of raising offspring. The explanations of Articles 8 and 12 of the ECHR and related case law reinforce the role of 'parenting'. Article 8 offers protection for 'private and family life'. The relevant case law shows that for family life to exist, a biological bond is not enough, but needs a further familial element [17]. Thereby there is a strong presumption of the intention to raise the child [18]. Article 12 is based on the idea that having children is the main source of founding a family [19]. The emphasis lies on begetting a child, or conceiving and consequently becoming a parent. The right to procreate, if there is one, means an entitlement to nurture children as a part of one's family, rather than fulfilling a desire to have biological offspring. The decision of the European Court of Human Rights (ECtHR) in *Dickson*, may be viewed as endorsing the ideas of family life and parents taking care of their descendants [20]. In this case, a prisoner and his wife were allowed access to fertility services. The Court argued that the welfare of the child could not hinder the applicants to seek access to treatment especially because the woman 'could have taken care of any child conceived' [21].

The traditional notion, notwithstanding its various guises, is characterized by its emphasis on the outcome of the reproductive process: the fact that a child will be born, and acquires much of its moral importance from the rearing aspect which it would entail. This has two implications.

First, the traditionally constructed right seems to cover a variety of events and processes, some of which are quite different in nature. In my view, four events and processes may legitimately be distinguished: the phase before conception, the period from conception to 'viability' of the fetus [22], the time from the onset of viability to the birth of the child, and the process of pregnancy, which I see as separate items on the list and which could be supported by the recognition of the 'right to manage one's pregnancy' [23, 25]. Different interests are involved in these different phases, and I discuss later on in the paper the problems that their amalgamation causes to the traditional notion. I maintain that, as the new notion has a very limited extent, and only applies to the first of these separate phases, it will overcome these problems.

Second, the stress on the outcome makes the interests of the eventual child an essential consideration. I will address below the complications created by the inclusion of this third party's good in what could – and as I argue, *should* – be an assessment of an individual's right to (be able to) produce progeny. I do not deny the existence of these interests, but hold them to be irrelevant for the content over which the right to procreate extends.

Where, in the following sections, I refer to the traditional notion of the right to procreate, I thus intend a right which aims at reproduction, and finds its moral relevance in the rearing of a child.

The Focus on Rearing a Child Fails to Recognize Other Important Elements of the Right

In its insistence on bringing up children, the traditional view fails to recognize other factors that have steered the approach towards reproductive legislation and policy making. For example, practices such as sterilizing people with intellectual disabilities and refusing IVF treatments on social grounds, do not necessarily prevent individuals from raising children, but have been condemned by an appeal to the right to procreate. Whereas they affect the opportunity to have genetic offspring they do not necessarily infringe the possibility of bringing up other people's children. It seems, then, that the rearing aspect cannot be the qualifying element of the right to procreate, although the traditional view in its purest form tends to ignore other considerations.

Three other aspects appear to define the right to procreate in the approach to the legislation and regulation. Firstly, a genetic link can be qualifying for the right to be at stake, as shown by the case of adoption. Adoption involves an intention to rear a child, yet it is uncommon to discuss this as an instance of the right to procreate. Secondly, a gestational link as such can justify claims to the right. While surrogacy arrangements sever the link between pregnancy and rearing, a woman's entitlement to offer these services has often been discussed as an instance of her right to procreate [13]. Thirdly, the capacity to procreate is sometimes all that is at stake when the right is evoked. Fertility treatments do not automatically lead to raising children, although they are typically demanded by appeals to the right to procreate. The right to treatment, if there is one, is a right to have one's procreative capacity activated or restored, not to bring up children. John Robertson states that 'the importance of procreation as a whole derives from the genetic, biological and social experiences that comprise it' [23, 25]. Whereas the traditional notion is dominated, for its moral relevance, by that last aspect, my alternative concept takes into account the other elements of this statement.

The Capacity to Procreate

This section suggests a new approach towards the right to procreate. It redefines the right to procreate according to the lowest common denominator; as a right to realize and control the capacity to procreate.

What is the Right to 'Realize and Control the Capacity to Procreate' [24, 26]

The right I envisage contends that everyone is entitled to retain, restore, realize and control his or her capacity to engage in the process of procreation; the process in order to conceiving or begetting offspring. That is to say that an individual has a right not to be impeded in attempts to realize his/her procreative capacity, included for the situation where he/she is affected by some dysfunction in this capacity, which medicine can remedy to make conception of a child possible. The emphasis of the notion lies on the 'capacity to procreate', by which I mean the ability to play an effective role in the process of procreation, with appropriate medical assistance if needed, and the condition that conception is not impossible. I do not claim that the conception or fertilization should be realized, or that an individual has right to a partner with whom to procreate, nor that he/she should be provide with complementary gametes. The right only endeavors to establish a physical state, which is 'natural' or comparable to it, and only aims to guarantee that conception could be possible [27]. Thus conceived, the new notion is not a right in which two persons are involved.

The following considerations demonstrate what I mean by the capacity to procreate. If a woman has no womb, she cannot conceive naturally, and so may be seen as lacking the 'capacity to procreate' as defined under the notion [28]. The new right implies that she should be allowed to establish a situation where conception is possible by, for example, a womb transplant or an external womb, or by paying for surrogacy. For infertile women and men, the right implies that they are allowed to seek to create the conditions under which they are in the same position as fertile individuals. They should be allowed to cure their infertility or to seek for 'fertile' gametes, so that fertilization in principle could be achieved. For a 'naturally' fertile person the notion implies that his capacity to procreate cannot be removed against his will. An individual should retain the physical ability required for the potential to actualize the mixing of his gametes with those of consenting others [29].

The revised notion focuses on the beginning of the reproductive process and has a limited scope. There are different phases in the reproductive process, but the revised notion only accounts for the conditions before conception and focuses on the ability to actualize the moment of fertilization. It does not deny that the other stages of the process are also relevant. To the extent that they are significant, other rights should be established to govern them. For example, the right to manage one's pregnancy could govern the process from the moment of conception until the birth of the child [30].

This focus on the beginning of the process places the emphasis on the existing individual and his needs. One implication of this is that the notion does not focus on the interests of the child, like many other approaches do, including the traditional one [31].

Where is the Moral Relevance in the Capacity to Procreate

What, then, is the moral relevance of the right to retain, restore, realize and control the capacity to procreate? Why would it be important for a person to have and to control this capacity, if this does not necessarily mean that he will have a child? In what way are an individual's interests served when the notion gives him this capacity to attempt to have offspring, but does not go further in the direction of becoming a parent? I consider three reasons why the notion is important as it is: for having a choice, for an individual's identity and as it is an original and basic condition for having a genetically related child [32].

Firstly, having the capacity to retain, restore, realize and control procreation means having a choice and this is important for human flourishing [33]. The capacity to procreate gives a person a choice either to enter or not to enter into the process of fertilization. He can engage in the process without hindrances, because the right has taken these away [34]. He is enabled to choose by his will, rather than by his physical state, what his engagement is. This is important; it makes the individual a 'free and dignified human being' and enables him to flourish as a 'maker of choices' [35]. Bonnie Steinbock holds that 'having the freedom to procreate if one chooses is something that everyone has reason to want' [8]. The importance of choice is also recognized by legislation on fertility treatments and case law on non-consensual sterilization. Legislation makes it possible to choose fertility treatments, even if having children cannot be guaranteed, and in the historical case law non-consensual sterilization was seen as a violation of the moral right to procreate, since it was irreversible and detrimental to future choice [36].

Secondly, the new notion endorses an individual's identity by recognizing the central role of genes, biology and natural instincts in procreation. John Robertson explains why genes are important for identity. He argues that 'it can be meaningful to say that one has reproduced when one has merely passed on genes and neither gestated nor reared the resulting child. A gene contributor may find genetic transfer a vital source of feelings connecting him or her with nature and future generations' [23, 25]. Similarly, Joseph Ellin argues that the special relation between humans and their genes can be seen in the preference for having one's own children over having children via adoption [37]. My capacity-centered notion, further, focuses on creating a biological starting point, as far as possible, and promotes that biological bonds can be established even when genes cannot be transferred, like in cases of donor gametes. Accordingly, my revised notion of the right to procreate forms a valuable instrument to this meaning of identity.

Thirdly, the capacity to procreate is an original and necessary condition for having a genetically related child. Without the capacity to procreate it is, in principle, not possible to have a genetically related child to rear. If having a genetically related child to bring up is important, then so is the necessary condition for making this happen [38]. This holds true, even though in some cases the child will not be genetically related but, for example, conceived by donor gametes; this particular case does not affect that having the capacity remains a prima facie condition, thus with general importance.

Could a Focus on the Capacity to Procreate Develop Into a Human Right?

In this section, I argue that since the revised and limited notion of the right to procreate can be universalized and since it is neutral, it can provide a better foundation for a human right than the traditional one. I have already argued that the notion is sufficiently important to form the basis of a right. This section analyses how this right could become a human right.

Universality is the first test that a human right has to pass. A right is universal, or can be universalized, if we can grasp and advocate the idea that it should be enjoyed by every human being at all times and in all places. Since this is easy to grasp and feasible to advocate in the case of realizing and controlling one's capacity to procreate, the revised notion passes the test. At least for people who are within a 'fertile age', the capacity to procreate seems to be something that we can picture for everyone to 'naturally' have if they so wish [39, 40]. The test is not, by comparison, so unequivocally passed by the traditional notion of producing a child or becoming a parent. The idea that everyone may become a parent or a care taker can be overridden, in many circumstances, by the interests of the child or even by those of the would-be parent or care taker. Accordingly, in the test of universalism, the revised notion is more promising as a potential human right than the traditional one.

Neutrality, as demonstrated by inclusivity and descriptiveness, is another element that makes the revised notion more adequate as a potential human right than the traditional concept.

Firstly, the notion provides virtually equal protection for both sexes and covers mutually those with a 'natural' capacity to procreate and those without it. A right to the capacity to procreate applies to men and women almost similarly [41]. Men and women have not completely similar positions, since in procreation there might be a 'short moment' during which the woman alone is in charge of the capacity; the time when the gametes are inside the woman's body but conception has not yet occurred [42]. The man at that phase will no longer have control, and the woman will decide over what happens with the gametes. Nevertheless, despite this time where the woman has ultimate control, the notion promotes far-reaching gender equality. Furthermore, since the model pertains to fertile individuals retaining the capacity and infertile individuals (re)gaining or attaining the capacity, it does not distinguish between these two groups. The right thus applies to everyone who is of a 'fertile age', which can be seen as an encouraging sign of the inclusiveness, essential for human rights [43]. The same is not true about the traditional notion, which is gender-biased and often understood as only protective of individuals who are fertile. It serves as a shield against interference, but does not necessarily offer anything for the 'infertile'. The revised notion goes further and overcomes this problem.

Secondly, since the revised notion focuses on individuals, it is less normative than the traditional one that placed value on having a family and on the outcome of the process. According to some feminist critiques, the emphasis on the value of the family makes women breeding machines – a woman who does not bear a child and form a family is not a true woman [13, 44–46]. The revised notion that no longer focuses on the outcome, but emphasizes the individual and his capacity, thus will not be subject to the same criticism.

Problems of Implementation and how These can be Solved by the New Approach

The implementation and codification of the right to procreate have been hindered by three elements in the traditional notion: the inclusion of the interests of the child; the vagueness of the right's scope; and its gender-bias. These problems arise as a result of the traditional emphasis on the outcome of the process; a child to rear, and the fact that it does not distinguish the phases preceding the birth of the child from this outcome, but covers all the way. The revised notion overcomes these problems by its well-defined focus, and stands a better chance of becoming a serious legal and human right.

The Interests of the Future Child Under the Traditional Notion

Since the traditional approach to right to procreate, as a right to rear, emphasizes the outcome of the reproductive process, it is necessarily concerned with the interests of the future child. Discussions about what these are, and how they can be best served have obscured the right to procreate, since the debates have had no incontestable outcome on how to take them into account. The revised notion deals with this problem by focusing on the existing individuals, in the phase before conception. The notion does not discard them, but provides that they should be considered at a different phase than to which the right extends.

Considering the interests of the child has provoked questions about how to define the right to procreate. Could such interests limit the right? For example, this has been suggested in controlling access to fertility treatments [48]. If a doctor suspects that the future child would be born in circumstances that were not in his interest, access could be limited [49]. Or, could the interests of the child influence the scope of the right. For example, it has been argued that the right to procreate should not extend to choosing the tissue of the child and creating a savior sibling, as this child would rather be born as an instrument than for its own sake. The right's scope could not extend to that practice [50].

There are various reasons why the interests of the future child need not be considered, and which justify a different focus. Firstly, the points that endorse why the right to retain, restore, realize and control the capacity to procreate is important make the interests of the child subordinate to them. Secondly, laws and regulations have more or less recognized that the child's interests cannot play a role as regards the pure act of natural procreation. It is accepted that a child can be taken away after birth, when it is born in 'harmful' circumstances [51]. Yet it is not accepted that, where this situation might arise, the capacity to procreate would be actively removed to prevent conception. Thirdly, considering the interests of the child is discriminatory, impossible and unnecessary. It is discriminatory because only individuals who need help with procreation are affected by the consideration of such interests [52]. They are screened in their demand for fertility treatments and their access to them is limited. Individuals who do not need help, on the contrary, can procreate without restrictions. Even licensing all parents, if anyone would be ready to resort to draconian measures like this, would not be a solution [53–55]. It would be impossible to establish the criteria for the license and it would not be practically possible to maintain these criteria in cases of natural procreation. Who could determine what a good parent is? A would-be parent might change his behavior after a child is born [56]. If a set of criteria were established, would this imply that fertile individuals who fail the test would be prohibited from procreating? What is in a future child's best interests? Is the answer a loving family, a wealthy family, one loving parent, or two parents whatever the status of their relationship? Furthermore, a detailed calculation of the future individual's interests is unnecessary anyway, as his moral interests are almost always better served by being born than by not being born at all, and his legal interests are non-existent and thereby irrelevant in this context [57, 58]. Although, for example, the Human Fertility and Embryology Act refers to the interest of the future child in section 13 (5), the section does not grant him any rights. It only assumes a moral status for it [59].

The revised definition focuses on existing individuals and forgoes the assessment of the interests of the future child. These interests are not relevant at the stage to which the right extends. It concentrates on the interests of those who are actually involved: it focuses on the beginning of the procreative process and deals only with the individual's capacity to procreate. This gives clarity to the new approach and allows it to form a better foundation for a universal human right. The notion does, however, not imply that the interests of the child are irrelevant, but the notion is not concerned with them.

The Unqualified Traditional Right is Too Vague and Too Controversial

Vagueness and ethical controversy characterize the traditional notion that entails the whole process from conception to birth. Although regulating practices at the beginning of life is always contentious, the traditional interpretation exacerbates controversy because its scope is too wide and too undefined.

Vagueness arises as to what the traditional right would cover. Is it a positive or a negative right? If were negative, it would rule out interference by the government. But does it also imply that individuals are not allowed to interfere with each other's activities? In that case the government would arguably have to provide protection against such interventions. Or is it equally a positive right? If so, then the government would be required to provide financial assistance for procreation. But how far does this requirement extend? A financially protected right to raise children seems impossible to secure.

Ethical controversy is equally widespread under the broad and unqualified notion, as each step in the process towards birth that it covers involves new issues. Problems range from the questions preceding conception to issues surrounding the birth of the child and beyond. Which interferences with gametes are ethically acceptable and when? Are screening gametes or modifying them among these? Can we legitimately meddle with the fertilized embryo in vitro, for instance, by Pre-Implantation Genetic Diagnosis (PGD) or sex selection? Which methods of reproduction would fall under the notion? Are cloning and Intra-Cytoplasmic Sperm Injection (ICSI) all right? Do we have a choice how to give birth? Can we choose to have a Caesarean section if we so desire? In addition, issues like family structure, sexuality and beliefs become relevant in considering the process from the beginning to the end [14]. The traditional notion, which does not distinguish the phases, accumulates an array of ethically controversial issues.

Controversy causes a problem for regulation because if consensus cannot be reached on an issue, legislators may justifiably be more reluctant to address this question at all. Especially on the human rights level, where agreement has to be reached internationally, things can be difficult. In sensitive matters like procreation, the difficulties are further amplified. Nation states may feel that issues like these are closely related to their 'identity', which means that they can be less open to compromises. The traditional approach to the right to procreate, with its contentious character, is not a good foundation for regulations at the level required for a human right.

The revised notion, on the other hand, reduces controversy and provides a clear basis for agreement [60]. As its scope is limited to the capacity to procreate, it involves fewer dilemmas that need to be resolved at the international level. For example, the issue of PGD would no longer fall under its scope, since it is not connected with the capacity to procreate. Nations could retain their ethical identity in issues like these, as they fall outside the revised scope. Furthermore, the new notion states clearly what it regulates, and provides a minimum level of protection for an essential part of the procreative process. Since it reduces detrimental ambiguity, the revised notion provides a better foundation for a potential human right than the traditional one [61].

The Traditional Notion is Dominated by Women

Women dominate under the traditional notion of the right to procreate, so much so that Margaret Brazier has referred to the right as a 'female right' [62]. The notion's focus is gender-biased, because under natural circumstances the woman's body is needed in the whole process from sexual intercourse to the birth of the child and because, according to the liberal contemporary Western reproductive ethos, she controls, or should control, the process. She can take contraceptives so that conception could not occur, or may abstain from taking contraceptives so that she can conceive. Fertilization will happen inside her body, the fetus will mature in her uterus and eventually the child will be born from it, so her bodily integrity will, by right, rule over the process from beginning to end [63]. She can claim ultimate decision-making capacity over what will happen with the fetus and she will dominate the procreative process.

The traditional entitlement does not recognize the man's role, and the ensuing gender bias makes it inadequate as a human right. Unless a man wrongfully interferes with the woman's bodily integrity, he has very little influence on the entire reproductive process. Under natural circumstances, his right to procreate can only extend to the decision of whether or not to engage in sexual intercourse for procreation [64]. In assisted reproduction, his influence can extend further, as evidenced by the case of *Evans* [65]. In this, the ECHR ruled that both partners had an equal right to become or not to become a parent. The decision was possible, since the frozen embryos that were produced in vitro by using the gametes of the two partners, were not yet placed inside the woman. Under English law, both parties would have to consent to the use of the embryo until that moment, and accordingly the decision-making powers of the man were momentarily extended [66]. However, these would again have been eroded after the implantation of the embryo. Although men had, historically, a greater influence on the process due to inequalities in marital and gender relations, this influence has now decreased in liberal Western legislatures, and the right to procreate has become a woman's right [67].

The revised notion, which focuses on the capacity of the individual to procreate, treats men and women almost equally, and this makes it better suited to be made a universal human right. Under this notion, both men and women have a nearly equal opportunity to retain, restore, realize and control the capacity, completely and independently of each other. Where it is argued that the burden between men and women will always be different in procreative activities, the capacity-based perspective of this paper cannot rebut this view, but relies on the fact that it adopts a starting point where the difference in gender burdens is reduced as far as possible. It could for example, be argued that even before conception, the stage that we are talking about in this paper, the prospect of begetting a child, may have a different psychological impact on a woman than on a man, or, equally, we could see how the 'scarce' availability of ova generates a different perspective on the reproductive burden, as compared to the 'wide' availability of semen. Since it seems difficult to imagine that at any other stage, the physical and the psychological involvement of men and women could be as equally footed as at the capacity-level, it can be said that the new notion does not favor one sex over another; it reduces gender inequality and it is a right for all humans.

Conclusion

This paper has sought to revise the right to procreate by suggesting a new notion that would deal with some of the problems of the traditional one. The paper has tried to show that under the traditional approach, where the right to procreate has been equated with the right to rear a child, the claims are muddled. It has suggested that the notion can be better understood by focusing on the capacity to procreate. This revised approach would overcome the obstacles to the traditional notion's establishment in law, created by the interests of the child, its vagueness and its gender biased focus. Furthermore, the revised notion that has been argued for has other characteristics, as identified above, which make it suitable to be classified as a human right.

Although the new notion does not provide an answer to all contentious and ambiguous problems, a right to realize his or her capacity to engage in the process of procreation may be a first step in creating a useful series of rights, to construe the concept of the right to procreate.

References

1. Wherever this paper refers to 'realization', this implies a realization of the physical prerequisites which provide the capacity to engage.
2. The paper will consistently refer to the right to procreate. Varying the term with the right to reproduce would not be correct since the two terms are not synonymous. Whereas the term procreation, in the definition of the right to realize and control the capacity refers to the 'technical' process in the beginning of the conception; including the process of fertilization, but also the conditions of the womb, the term reproduction has already a relationship with the outcome of the pregnancy: reproducing infers a child.
3. G. Pennings. Personal desires of patients and social obligations of geneticists: applying preimplantation genetic diagnosis for non-medical sex selection. *Prenat Diagn* 2002; 22: 1123–1129: 1128.
4. G.d. Wert, et al. *Ethics and Genetics: A workbook for Practitioners and Students*. New York, Oxford: Berghahn, 2003.

5. S. Millns. Reproducing inequalities: assisted conception and the challenge of legal pluralism. *J Soc Welfare Fam Law* 2002; 24: 19–36: 26.
6. A. Halpin. *Rights and law, analysis and theory*. Oxford: Hart publishing: 284. 211, 1997.
7. The moral importance is understood here as the decisive factor in appealing for the right to procreate: the element that makes the right utmost important.
8. B. Steinbock. Rethinking the Right to Reproduce. In *Harvard Centre for Population and Development Studies*. University at Albany/SUNY, 1998.
9. 'He' in this paper may be substituted by 'she', equally for him/his and her.
10. O. O'Neill. *Autonomy and Trust in Bioetichs*. 2002 edn. Cambridge: Cambridge University Press, 65, 2002.
11. O'Neill. Begetting, Bearing and Rearing. In *Having Children: Philosophical and Legal Reflections on Parenthood 25*. O. O'Neill & W. Ruddick (eds.) New York: Oxford University Press, 25–38, 1979.
12. Using the right to procreate synonymous to the right to procreate is not uncommon. See for example D. Statman. The Right to Parenthood. *Ethical Perspect* 2003; 10: 224–235.
13. T. Bayne and A. Kolers. 2006. Parenthood and Procreation. Stanford Encyclopaedia.
14. J. Boivin and G. Pennings. Parenthood should be regarded as a right. *Arch Dis Child* 2005; 90(8): 784.
15. O'Neill's elaboration on the right to procreative autonomy of 2002, seems not to deviate much from her vision in the earlier work of 1979.
16. E.S. Scott. Sterilization of Mentally Retarded Persons: Reproductive Rights & Family Privacy. *Duke Law J* 1986; 5: 806–865: 829.
17. *Lebbink v. the Netherlands* [2004] *(45582/99)* European Court of Human Rights Para. 37–41.
18. *Keegan v. Ireland* [1994] *(16969/90)* European Court of Human Rights Para. 43.
19. L.E. Kalkman-Bogerd. Artikel 8 en 12 in verband met kunstmatige voortplanting en draagmoederschap. In *Het EVRM en de gezondheidszorg*. J.K.M. Gevers (ed.) Nijmegen: Ars Aequi Libri: 122, 1994.
20. *Dickson vs The UK* [2007] *(44362/04)* European Court of Human Rights (Grand Chamber).
21. See para. 76 of the judgment
22. It is submitted here that the viability of the fetus is a problematic concept. The legal and medical professions use a different definition, and advances in technology continue to change the meaning of the notion as well as the point at which the fetus can survive.
23. The notion of a right to mange pregnancy is borrowed from work of John Robertson. J.A. Robertson. Procreative liberty and the control of conception, pregnancy, and childbirth. *Va Law Rev* 1983; 69: 405–465: 408. It is distinguished from the two other phases where conception has already occurred. An elaboration on the right to manage one's pregnancy and an explanation of how the two other events equally have their own rationale, is destined for future work.
24. R. Cook, et al. *Reproductive Health and Human Rights*. Oxford: Oxford University Press, 3–554, 2003.
25. Robertson. 408. *Op. Cit. note. 22*
26. Due to the limited space, this paper will only deal with the notion's onset and will not consider its limits. A subsequent paper will elaborate further upon the new notion.
27. It is controversial to use a like 'natural' state to refer to the physical state of a person who is fertile; a person may be born being infertile and arguably his 'natural' state is thus being infertile. This paper will use terms like 'natural' nevertheless, but will use the term to refer to the physical state of a majority. It will use inverted commas to acknowledge to the ambiguity of the term.
28. I do not mean to argue that a man should be provided with a womb. This would fall outside of the capacity 'naturally' existing in the majority of men. See note 27.
29. Or the donor gametes that he can 'use'.
30. This notion is borrowed from John Robertson *Op. cit. 23.*
31. See section IV.1 for further elaboration on this issue.

32. Even if donor gametes are used and no genetic material of the would-be parent would be transferred, this arguments holds its value as it creates an ability; see further on in the section.
33. This choice extends to the capacity to engage in the process of fertilization. It is submitted here that this choice will not necessarily imply that one conceives successfully, but it is merely a choice that arises from the fact of an ability to engage than being hindered by a disability.
34. There might be cases where individuals will resort to donor gametes since they have no gametes of themselves, and according to the notion would not have the capacity to procreate. In that case the individual will not be able to mix 'his' gametes, but he will be able to mix the donor's gametes.
35. M.C. Nussbaum. 1999. *Sex and Social Justice.* New York: Oxford University Press, 3–476: 46.
36. See judgment of Heilbron J in *Re D* [1976] (*1 All ER 326)* Family division the more recent case law provides no clear statement either on how the right to procreate should be perceived and non-consensual sterilization still takes place. See for example: *Re Z (Medical Treatment: hysterectomy)* [2000] *1 F.L.R 523, A National Health Trust v. C* (2000, unreported but available via Lexis).
37. J. Ellin. 1978. *Contemporary Issues in Biomedical Ethics.* Clifton, New Jersey: The Human Press Inc. 135–141: 135.
38. Why conceiving a genetically related child could be perceived as important, is set out above, and summarized by Robertson. 409 O*p. cit.* note 23
39. The larger part of the population is born having a 'natural' capacity to procreate, so it seems obvious to picture this for everyone. If the notion would extend to those individuals outside the 'fertile' age range, this would create problems to universalize the notion. Various arguments such as the interests of the child, health reasons of the would-be parent, distributive justice, would argue against why this is not universally desirable. See also note 40
40. G. Pennings. Postmenopausal Women and the Right of Access to Oocyte Donation. *Journal of applied philosophy* 2001; 18: 171–181.and D. Cutas. Postmenopausal Motherhood: Immoral, Illegal? A Case Study. *Bioethics* 2007; 21: 458–463.
41. See section IV.3 for a further explanation.
42. For example, in natural procreation this situation arises from the moment that the man has ejaculated to the moment of conception. Although this process may take 8–10 days after the ovulation, this is still considered as a 'short' moment in respect of the entire scope of the notion and the pregnancy. From the moment of conception, the revised notion no longer applies.
43. I acknowledge that the notion of 'fertile' age is not clear-cut. Not only for elderly women or men it may be difficult to establish what their 'fertile' age is, this may equally be a problem for young individuals: when are they of fertile age and when is their absence of fertility abnormal? It needs elaboration in another paper on how to deal with this obscurity.
44. I. Aristarkhova. Ectogenisis and Mother as Machine. *Body and Society* 2005; 11: 43–59.
45. R.A. Fenton. Catholic doctrine versus women's rights: the new Italian law on assisted reproduction. *Med Law Rev* 2006; 14: 73–107.
46. L.M. Purdy. What Feminism Can Do for Bioethics. *Health Care Anal* 2001; 9: 117–132.
47. B. Berg. 1995. Listening to the Voices of the Infertile. In *Reproduction, Ethics and the Law: Feminist Perspectives.* J.C. Callahan (ed.) Bloomington Indianapolis: Indiana University Press.
48. E. Jackson. Conception and the Irrelevance of the Welfare Principle. *Mod Law Rev* 2002; 65: 176–203.
49. See for example article 13 (5) of the Human Fertility and Embryology Act. Under the new guidelines there is a presumption of treatment H.F.a.E. Act. 2006. Review of the Human Fertilisation and Embryology Act. command of Her Majesty edn: Secretary of State for Health.
50. C. Berry and J. Engel, *Saviour Siblings* Available at:http://www.cmf.org.uk/literature/content.asp?context=article&id=1317 [Accessed 22 November 2007].
51. See for the English legislation section 31 of the Children Act 1989.
52. Even if it would be argued that restricting reproduction for the interests of the child would not necessarily be discriminatory in a moral sense; the restrictions would then also apply to

natural reproduction, such restrictions remain practically impossible, so that the restrictions will have practically a discriminatory effect.

53. H. LaFollette. Licensing Parents. *Philosophy and Public Affairs* 1980; 9: 182–197.
54. L.E. Frisch. On Licentious Licensing: A Reply to Hugh LaFollette. *Philosophy and Public Affairs* 1982; 11: 173–180.
55. R.E. Redding. The impossibility of parental licensure. *Am Psychol* 2002; 57: 987–988.
56. Hereby I distinguish procreation from adoption. I hold it impossible to assess a parent's quality in natural procreation, since the child is not yet born. The parent must therefore be approached with a presumption of innocence since the child's birth might change the would-be parent completely. This presumption would count for each child on its own, so that even though a parent has neglected children in the past, this should not be a reason to infringe with his/her right to control or realize the capacity.
57. The claim that the future individual's interests are almost always better served by being born, is open for discussion. The limited extent of this paper, does, however, not allow to deal with this discussion. For an elaboration on the issue is referred to D. Benatar. 2006. *Better Never to Have Been: The Harm of Coming into Existence*. Oxford: Clarendon Press, 237.
58. C.B. Cohen. 1997. The Morality of Knowingly Conceiving Children with Serious Conditions: An Expanded 'Wrongful Life' Standard. In *Contingent Future Persons. On the Ethics of Deciding Who Will Live, or Not, in the Future.* J.C. Heller and N. Fotion (eds.) New York: Springer-Verlag
59. Section 13 (5) is currently under revision. This revision seems, however, not to have any implications for the assumed status of the future child, but only how its interests should be met.
60. The whole of paragraph IV sets out which controversial elements of the traditional notion the new notion overcomes.
61. The revised notion brings other issues about its limitations and boundaries that need full consideration. So may be argued that the new notion is too narrow. The limited amount of words of this paper does not allow elaborating on these issues and so these considerations will have to be reserved for another paper. This paper will just be concerned with explaining why the traditional notion is inadequate.
62. M. Brazier. 1998. Reproductive Rights: Feminism or Patriarchy. In *The Future of Human Reproduction.* J. Harris and S. Holm (eds.) Oxford: Oxford University Press: Clarendon Press Margaret Brazier seems, however, not to endorse that the right should be purely a right for women, but only refers to it to illustrate its gender-bias.
63. Under the current stage of technology, artificial wombs are not yet available to bear the child.
64. A. Grubb & D. Pearl. Sterilisation and the Courts. *C LJ* 1987; 46: 439.
65. *Evans vs the UK* [2007] *(6339/05)* European Court of Human Rights (Grand Chamber).
66. The 'use' can be equated in this case with placing the embryo inside the womb.
67. Although in English law (female) sterilization never formally required the consent of both partners, it was common practice not to operate without consent of her husband. In certain religious convictions there still is a presumption in favor of consent of both partners for sterilization.

Part II
ART – Mind the Gap

Chapter 4
'Can't Avoid It, Can't Afford It': Assisted Reproduction in Israel and Bulgaria

Frida Simonstein and Ekaterina Balabanova

Introduction

Søren Holm observes n Chapter 2 that the effective freedom of a full exploitation of the new reproductive opportunities is only available to those who are in a position to pay privately. And, that the increase of procreative liberty created by the developments of ARTs has also indirectly decreased the effective freedom of some people; because the broad social acceptance of ART in many countries has changed social default expectations with regard to reproductive choices. This chapter offers an empirical study of these observations.

Reproduction is recognized as a Human Right; however the negative population growth in Europe suggests that people are not keen to use this right [1–4]. Some countries are responding to the demographic challenge of negative growth by the implementation of policies which could encourage women to have more children. Policies (religious and/or governmental) affecting reproduction are nothing new. The first policy put in writing affecting human reproduction, can already be found in the Biblical command 'be fruitful and replenish the Earth' [5]. For their part, governments have always had a say on reproductive matters [6].[1] In fact, reproductive policies can be found everywhere, all over the world. For example, China has implemented restrictions on the number of children that each couple can have [7]. Less known, perhaps, is that 80 per cent of all countries still prohibit abortion in some way; and in those countries where abortion is permitted, some restrictions still apply.[2]

Assisted reproduction treatments (ARTs), such as IVF (In-Vitro Fertilization), have been added to the list of topics in which governments may exercise policies affecting reproductive choices. It is estimated that between 10 and15 per cent of

F. Simonstein (✉)
Department of Health Systems Management, Yezreel Valley College, Israel
e-mail: fridas@yvc.ac.il

[1] Foucault would say that this concern arose at a specific moment – 18th century – when the 'population' was created as an object of policy [see 36].

[2] From these countries 36 would not make an exception even to save a woman's life [for more see 37].

F. Simonstein (ed.), *Reprogen-ethics and the Future of Gender*, International Library of Ethics, Law, and the New Medicine 43, DOI 10.1007/978-90-481-2475-6_5,

all couples will face some form of infertility; and in fact, assisted reproduction, initially aimed at helping 'desperate' women to have babies, has long since become a burgeoning industry. Since the birth of Louise Brown, the first 'test-tube' baby in 1978, assisted reproduction has blossomed [8]. By 2006 it was reported that 3 million babies were born with the aid of IVF [9].

As with any other reproductive policy, assisted reproduction affects the whole population, but particularly women. The level and quantity of assisted reproduction, and policies regarding it, however, differ sharply in different parts of the world. Differences sometimes are so substantial that a comparison between countries could be considered unfeasible. On the other hand, in order to get a better perspective in a world that is rapidly becoming a global market, making an assessment of similarities and distinctions regarding reproductive policies in different countries, and their effects on women, is important.

Bulgaria and Israel, for instance, share the same size of population, are markedly paternalistic and both have strong pro-natalist cultures, by which large families are expected by society. Contrasting sharply, however, Israel and Bulgaria differ radically on the number of children actually born. This paper is a comparative survey of assisted reproduction in Bulgaria and Israel and results from friendly and long lasting discussions between two scholars from these countries. The aim of this paper is to present data on women's status in Bulgaria and Israel and to assess both states' policies and involvement in assisted reproduction. The paper examines how assisted reproduction, as reflected by present policies in both countries, influences women's reproductive choices and also considers whether assisted reproduction makes women 'better off'.

Starting Point

On the surface, a comparative look at Bulgaria and Israel might be quite surprising. In reality, however, the two countries share a number of similarities that make this comparison fruitful. Table 4.1 shows that both countries are of similar size – Bulgaria's population is 7 679 290 (2006) and Israel's – 7 150 000 (2007). The number of women in Israel and Bulgaria is greater than the number of men. The ratio between women and men is 103:100 and 106:100, respectively. The average age of women's first marriage in Israel is 24.2 and in Bulgaria 25.5 [10–12]. In both countries women get paid less than men for doing the same job – women's salaries are on average 26 per cent lower than men's. Women are also primarily engaged in

Table 4.1 Bulgaria and Israel: similarities

	Bulgaria	Israel
Population	7 679 290	7 150 000
Women: Men	106:100	103:100
Average age of first marriage	25.5	24.2
Number of children per woman	1.3	2.9

professions that are traditionally perceived as 'female' such as nurseries, primary schools, secretaries, cleaners, etc. Very often pregnancy and motherhood are reasons for hidden discrimination, although in both countries there are laws protecting women [13]. However, despite these significant similarities, when it comes to the fertility rate it could not be more different.[3] Israel boasts an average of 2.9 children per woman compared with the modest 1.3 in the case of Bulgaria [10, 11].

Still, both countries perceive that they face a demographic threat (although caused and felt differently). While it could be important to see how this 'problem' is produced in the first place, the focus of this paper is on how Bulgaria and Israel deals with it as reflected in their policies for assisted reproduction.

ARTs in Bulgaria

In traditional discourses childless women in Bulgaria are described as 'defective', and also as dangerous and to be avoided [14]. A woman who could not conceive was often marginalized, seen as both powerful to cause harm but also passive and desperate, 'at the mercy of society and healers'. Some traces of this view can be discovered even in the present day. While the explicit ostracization of childless women has become significantly tempered, stigmatization in more subtle forms does still exist.[4] It happens mainly through identifying 'true' women as mothers, through explicit and implicit blame directed solely at the woman and through a symbolic exclusion and self-exclusion from social life and family identity revolving around children [14, 15]. In addition, presently women who postpone having children for educational or professional reasons tend to be portrayed as selfish and irresponsible and if difficulties with conception appear, the attitude is that these are deserved. Considering this pressure, assisted reproduction is an important option for women facing infertility.

The first IVF baby in Bulgaria was born in 1988 and the girl, Iliana, was the first to be born by this method in the Balkans. It is estimated that since then around 1000 babies have been born as a result of IVF [16]. Currently 230 000 couples in Bulgaria cannot have children because of fertility problems [17]. This figure could be even higher as a survey conducted for the UN Fund on Population in 2003 registered 270 000 couples with damaged reproductive health [18]. This represents around 15 per cent of all Bulgarians at childbearing age. IVF treatments are performed in 15 clinics and cost from \$2 830 to \$6 360,[5] which compares to the average monthly salary in the country in 2006 of \$274 [19]. In 2004 IVF was included for the first time in the range of health services covered by the National Health Insurance Fund (NHIF) [20]. The NHIF funded the treatment of around 1000 women who

---

[3] Of course, there are other important distinctions such as comparative wealth, GDP and its distribution, income stratification, inequality, etc., however they are beyond the scope of this paper.

[4] Strong pronatalism was formally endorsed in socialist times, when stimulating birth in Bulgaria was an official government policy [for more on this see 15].

[5] The figures are calculated at the exchange rate of 1 USD = 1.41506 BGN (19/07/2007).

were between 18 and 40 years old and (depending on clinical requirements) it paid between $706 and $2 119 per woman. In 2005 the NHIF covered treatment of 1270 women. In 2006 the number increased to 1500 women, before dropping again in 2007, when the figure reached 1200 patients. It is worth noting that the NHIF covers only the hormonal medicines – the preceding tests and actual procedure are covered by the patient [21]. Also, according to regulations passed on 9 January 2006, a woman can have the treatment only once under the NHIF [22].

The limited existing support for reproductive technology is paralleled by a lack of subsidies for contraceptives and the acceptance of abortions. Abortions are legal in Bulgaria and the decision to have an abortion is seen as a civil right, removed from the moral or religious domain.[6] In fact, induced abortions are seen as the prevailing birth control method in the country [23] and are additionally stimulated by a private medical practice that profits from them [14]. The number of abortions in 2006 was 67 000 and only 4 per cent of women in childbearing age were using contraceptives [24].

ARTs in Israel

Contrasting sharply to the state of ARTs in Bulgaria, by the mid 90s there were already more fertility clinics per capita in Israel than in any other country in the world (22 units for a population of 5.5 million) – four times the number per capita in the US [25] and by 2003 the number of fertility clinics had increased to 24 [26]. Reproductive technologies in Israel are subsidized as a standard part of the basic basket of health services by Israeli national insurance. Every Israeli, regardless of religion and marital status is eligible for unlimited rounds of IVF treatment free of charge, up to the birth of two children – even if the woman has children already [27]. In contrast to this extraordinary level of support for reproductive technology, treatments that limit family size remain unsubsidized in Israel. Family planning services do not receive state support and are funded only on a charitable basis. Contraceptives are not part of the basic basket of medical services covered by Israeli health insurance, and abortion is illegal (unless the pregnancy threatens women's life, it results from rape or adultery, the mother is underage, or the fetus has a major malformation or illness). In addition, in 1996 the Israeli Knesset (parliament) passed the Embryonic Carrying Agreement Law, making Israel the first country in the world to legalize surrogate mother agreements, provided they are regulated by a government-appointed commission [28].

This is probably not too surprising considering that Jewish women in Israel (but also elsewhere) are under extraordinary pressure to reproduce, whether they are married or not. The barren woman is an archetype of suffering in the Israeli-Jewish mentality. From the childlessness of the matriarchs in the Book of Genesis (about

---

[6] Periods of limited abortion rights existed in Bulgaria during the socialist time as part of the official policy of stimulating child-birth [see 15].

which every Israeli schoolchild learns from the age of six) Israelis are taught that barrenness is tragic for a woman. The childless woman is pitied; childbirth is a female success [25]. Childlessness is also tragic for a Jewish man who is commanded to procreate according to Jewish tradition. Reproduction is also reinforced by demographic and political beliefs. Cultural and demographic 'needs' have in turn produced a burgeoning business of IVF technologies, which combined result in an aggressive marketing of pro-natality and IVF in Israel [29].

IVF Performance

A useful perspective can be obtained from the following two tables that illustrate how Bulgaria and Israel are performing with regard to the IVF treatment. Although the data is for 2002, it shows well the disparities between the two countries. While Bulgaria had only 10 IVF clinics in 2002 with every clinic 'covering' a population of roughly 1 million, Israel had 23 clinics whereby every clinic covered a population of 280 000 (Table 4.2). (As mentioned before, the number of clinics has gone up in both countries, but the rise in the number in Bulgaria is markedly more substantial – a 50 per cent increase).

The difference is significant also in terms of the number of treatments in both countries. In Bulgaria 600 treatments were performed during 2002, which equalled 75 treatments per 1 million population. By contrast, in Israel 15 000 treatments were carried out i.e. 2 500 treatments per 1 million population (Table 4.3).

Table 4.2 Number of IVF clinics in Europe 2002 [30]

Country	Number IVF clinics	Population per IVF clinic
Bulgaria	10	~1 000 000
The Netherlands	20	490 000
Sweden	12	740 000
United Kingdom	76	770 000
Czech Republic	18	580 000
Hungary	10	1 000 000
Slovenia	3	660 000
Israel	23	280 000

Table 4.3 Number of treatments per year in Europe 2002 [30]

Country	Number of treatments	Number of treatments per 1 million population
Bulgaria	600	75
The Netherlands	9 800	650
Sweden	6 012	680
United Kingdom	29 185	500
Czech Republic	7 000	660
Hungary	2 800	280
Slovenia	1 331	675
Israel	15 000	2 500

Some Effects in Israel – Too Much of a Good Thing?

Fertility specialists in Israel have emerged as global leaders in the area of research and development of new reproductive technologies making Israel a regional hub for infertility treatment attracting patients from all over the Middle East and Europe.[7] Moreover, it is unlikely that a physician would suggest to a woman undergoing IVF to 'consider giving up and carrying on with her life'. Women who have succeeded in having a baby by IVF in Israel (happily for them) report that this suggestion was never made [31]. On the contrary, the advice they received was to 'keep trying' and for these women this advice certainly turned out to be right. Yet while IVF pictures show happy smiling women holding a newborn, statistically there are significantly more women who do not succeed in having a baby through IVF [4]. For these women, the very 'generous' policy for unlimited IVF cycles in Israel may turn out to be a never-ending nightmare.

The 'culture of perseverance' that has developed with respect to IVF (worldwide) is well known [32]. But this is a 'culture' also highly encouraged by IVF clinics. Less advertised is the fact that the vast majority of women undergoing IVF will *never* conceive [4].[8] Thus, the culture of perseverance, warmly advocated by IVF clinics, may work for some women, but it will not work for most. As a result, a number of women can get trapped in IVF treatment for many years. Reportedly, this happens worldwide [32]. However, in Israel, because the number of IVF cycles covered by the Health Funds is unlimited, there are no financial disincentives affecting the continuation of treatment, and this further deepens the trap of *endless* infertility treatments [31].

Women may give up at some point either because they cannot stand the pain of egg retrieval, or the bad moods linked to hormonal excess and/or the stress of the treatment. Yet these women in Israel – in most cases – remain silent [31]. Not surprisingly, since this environment of 'friendliness' to women's right to become mothers encourages women to keep trying to become pregnant. Thus, allowing a woman to go through as many IVF cycles as she can handle (no matter the consequences in terms of work, family, health) [31]. Moreover, if a woman decides to quit from treatment under these 'ideal' conditions, she shoulders the blame for the failure.

Some Effects in Bulgaria – Not Enough of a Good Thing?

Fertility treatment in Bulgaria is obviously not adequate for the needs of the country.[9] The high price of treatments, together with the difficult financial situation of

[7] This has naturally developed as business which generates further incomes both for the clinics and the state.

[8] A study conducted at Yale University in 1992 looked at 579 women who overall underwent 1257 cycles of IVF. Only 129 viable pregnancies occurred [4].

[9] It is interesting to note here what happens with adoption, the possible alternative to assisted reproduction. In Israel the number of people wanting to adopt exceeds the number of children waiting for adoption. In Bulgaria, it appears that it is foreigners who are more interested in adopting Bulgarian children then the Bulgarians themselves.

many families and the lack of state involvement, offer a straightforward explanation of why this is the case. As already indicated above, until 2004 neither the Health Ministry, nor the NHIF (or any other political or public organizations) were involved in financially assisting the treatment of infertility. When financial involvement finally happened it was done only partially and for a very small number of women. It is quite obvious that with provisions like these the whole notion of IVF treatment in Bulgaria significantly differs from the way this is done in most developed countries. Even the terminology differs – while most countries talk about the treatment of *infertility*, in Bulgaria the treatment is for *sterility*, a term with greater negative connotations.

Many childless couples report that the approach towards them in Bulgaria is completely commercial. This most likely is typical of the health care in the post-communist era, especially considering that the clinics are owned by the doctors themselves. The IVF treatment is very often undertaken without first trying to establish the reasons for infertility, hence rendering many procedures fruitless [33]. Problematic are also the number of couples to be treated, the choice of doctors/clinics that get to conduct the treatment, the exact choice of medicines (and their prices) as well as the qualifications of the doctors carrying out the procedures [34]. Limiting access of NHIF partially-funded IVF treatment to 1200 patients as is the case in 2007 leaves a number of insoluble dilemmas: Why was the number set at 1200? How and by whom was the selection of the 1200 women carried out? In addition, there are questions regarding the manner in which the NHIF as an institution determines which doctors/clinics will get the 1200 patients. It is argued that IVF treatments are currently monopolized by certain doctors and clinics, for example [21, 34]. The patients are distributed between a relatively small number, and this appears to happen regardless of clinical success rates. No independent checks are being conducted, and the professional qualities and experience of some of the doctors performing the procedures are questionable. Vatev [34] outlines one more problem – the issue of medicines being used. It is a practice in IVF clinics around the world to use a range of medicines and choose among them the most appropriate for the individual patient. It appears that in Bulgaria the medicines included in the programme funded by the NHIF are generally among the most expensive in the world. If this is indeed the case then the question of commercialization at the expense of patients' interests comes to the forefront – a better policy in this respect might mean a higher number of childless couples could be given the chance of becoming parents.

The focus on the access to the technology, however, has completely diverted attention from the question of the actual experiences of the women undergoing the IVF treatment. In sharp contrast to Israel no studies and research have been carried out exploring the role of the clinics performing IVF, the messages their give to women and the perceptions and the feelings of the women involved in the process. Any discussions of the IVF in the public space centre on the financial aspect of the treatment and this appears to be the main concern for the women. In reality, the cost of the treatment has the effect of preventing women from keeping on trying to get pregnant and feeling trapped as is the case in Israel.

Some Conclusions

In both countries, it is difficult to obtain detailed information about the real situation of reproductive health. In Bulgaria, that which is available is at times contradictory and most of the times incomplete. In Israel, the focus is normally on successes at the expense of any negative effects on women. In Bulgaria, there is little discussion of demographic problems in the public sphere and even less action to address them. By contrast in Israel the public sphere is filled with discussions of demographic problems (legacy of Holocaust) and any action to address this is seen as appropriate. Reproductive policy in this country has been presented as the state's concern for women's rights to become mothers; yet since other women's rights are inexistent (i.e. abortion) the right to motherhood becomes a convenient cover-up for a policy that in fact aims at increasing natality for demographic reasons. These policies are further promoted by the health care system and women may become instrumental in this scenario.[10]

This study has shown two extremes of assisted reproduction policies. In both cases it is crucial to identify the problems of women for the sake of those women involved since in both instances it appears that their welfare is being compromised. Moreover, the Israeli scenario is taken as an example to be followed worldwide because it is regarded as being very generous towards women. Nonetheless, it comes to the point where women in this situation simply cannot avoid using the existing technology, thus ending up worse-off in too many cases. On the other hand, most Bulgarian women cannot afford the technology at all and have to fight for access to it, which in effect removes the discussions surrounding the IVF treatment from its effects on women and their experiences of it.

Clearly, both approaches need to be modified in order to meet not only the demographic needs of Bulgaria and Israel, but also the needs of the women there. Moreover, in a global world it is difficult to predict the exact consequences of such disparate policies. Reproductive tourism and the world trade in eggs (for both reproductive and research purposes) are two emerging examples of the global effects of national policies [35]. Undoubtedly, more in-depth studies on ART policies and implementations in different countries and their actual consequences are needed.

References

1. UN. 1948. *Universal Declaration of Human Rights*. Article 16: Para 1 (available online at http://www.un.org/Overview/rights.html)
2. CBS. 2003. 'European Birth Rate Declines', CBS News 27 March (available online at http://www.cbsnews.com/stories/2003/03/27/world/main546441.shtml)
3. EU. 2005. 'Europe's changing population structure and its impact on relations between the generations', EU Press Release 17 March (available online at http://europa.eu/rapid/pressReleasesAction.do?reference=MEMO/05/96&format=HTML&aged=0&language=EN&guiLanguage=en)

[10] Although women sign an informed consent form, this is a formality which aims at protecting physicians.

4. Kaplan, E.H., A. Hershlag, A. H. DeCherney and G. Lavy. 1992. 'To be or not to be? That is conception! Managing in vitro fertilization programs', *Management Science* 38(9): 1217–1229
5. *The Bible: Genesis*. King James Translation.
6. Solinger, R.2005. *Pregnancy and Power. A Short History of Reproductive Politics in America.* New York: New York University Press
7. People's Republic of China. 2001. *Population and Family Planning Law of the People's Republic of China.* (available online at http://english.gov.cn/laws/2005-10/11/content_75954.htm)
8. Henig, M.R. 2004. *Pandora's Baby.* New York: Houghton Mifflin
9. Horsey, K. 2006. 'Three million IVF babies born worldwide', *Progress Educational Trust* 28 June (available online at http://www.ivf.net/ivf/index.php?page=out&id=2105)
10. NSI. 2006a. *Population*. Sofia: National Statistical Institute. (available online at http://www.nsi.bg/Population/Population06.htm)
11. Israel Central Bureau of Statistics. 2005. *Israel in Figures*. (available online at http://www.cbs.gov.il)
12. Jewish Virtual Library. 2007. *Latest Population Figures for Israel*. (available online at http://www.jewishvirtuallibrary.org/jsource/Society_&_Culture/newpop.html)
13. CITUB. 2006. *Women in Bulgaria*. Sofia: Confederation of Independent Trade Unions in Bulgaria
14. Todorova, I. and T. Kotzeva. 2003. 'Social discourses, women's resistive voices: facing involuntary childlessness in Bulgaria', *Women's Studies International Forum* 26(2): 139–151
15. Todorova, I. and T. Kotzeva. 2006. 'Contextual shifts in Bulgarian women's identity in the face of infertility', *Psychology and Health* 21(1): 123–141
16. Dimitrov, R. 2004. 'The first IVF baby in the Balkans was born 16 years ago in Bulgaria', *Доктор* 26 January (available online at http://www.zachatie.org/modules.php?name=News&file=article&sid=141)
17. Tencheva, D. 2006. 'A month devoted to the fight against infertility starts: 230,000 families are childless', *Trud* 19 October
18. UN. 2003. *Survey of the Population Fund* (available online at http://www.zachatie.org/modules.php?name=News&file=article&sid=175)
19. NSI. 2006b. *Labour Market*. Sofia: National Statistical Institute. (available online at http://www.nsi.bg/Labour/Labour.htm)
20. Zdravkova, I. 2003. 'The Health Fund pays for IVF next year', *Sega* 17 December
21. Damianov, L. 2007. *Reproductive Health in Bulgaria 2007*. Report at the Conference on the Reproductive Health, Sofia, 16 June. (available online at: http://www.zachatie.org/index.php?option=com_content&task=view&id=825&Itemid=23)
22. Davidova, F. 2006. 'Five outrages following a hidden official document', *Praven Sviat* 5 April
23. Mirchev, M. 1996. Reproductive Behaviour, Family Planning, Use of Contraceptives. Agency for Social Surveys and Analysis, Sofia
24. Vojnova, L. 2007. 'Only 4% of the Bulgarian women take contraceptives', *Dnevnik* 13 June
25. Kahn, S.M. 2000. *Reproducing Jews: A Cultural Account of Assisted Conception in Israel.* Durham, NC: Duke University Press
26. Shalev, C. and S. Goldin. 2006. 'The uses and misuses of in vitro fertilization in Israel: some sociological and ethical considerations'. *Nashim: A Journal of Jewish Women's Studies & Gender Issues* 12: 151–176
27. National Health Insurance Law. 1994
28. Surrogate Mother Agreements (Approval of the Agreement and Status of the Child). 1996. *Law*, SH 1577: 176
29. Gross, M. L. and V. Ravitsky. 2003. 'Israel: Bioethics in a Jewish-Democratic State', *Cambridge Quarterly of Healthcare Ethics* 12: 247–255
30. Damianov, L. 2002. *A possibility to prevent the catastrophe* (available online at http://www.zachatie.org/modules.php?name=News&file=article&sid=116)

31. Remennick, L. 2000. 'Childless in the land of imperative motherhood: Stigma and coping among infertile Israeli women', *Sex Roles* 43:821–841
32. Kalbian, A. 2005. 'Narrative ARTifice and Women's Agency' *Bioethics* 19: 93–111.
33. Docheva, D. 2006. 'Women protest against the one-time financing of the IVF treatment', *BG-Sever* 22 March
34. Vatev, I. 2006. *Report on the 'Treatment of women's infertility' programme for 2005* (available online at http://www.zachatie.org/modules.php?name=News&file=article&sid=264)
35. Waldby, C. 2007. 'Oöcyte markets: women's reproductive labour in embryonic stem cell research', *New Genetics and Society* 3
36. Foucault, M. 1990. *History of Sexuality: An Introduction.* New York: Random House
37. Centre for Reproductive Rights. 2007. *The World's Abortion Laws* (available at http://www.reproductiverights.org/pdf/pub_fac_abortionlaws.pdf)

Chapter 5
Assisted Reproduction in Developing Countries: The Debate at a Turning Point

Effy Vayena

Introduction

A recent report in the journal Nature described the story of a woman named Betty Chishava from Harare, Zimbabwe who was thrown out of her family home because she failed to conceive and refused to sleep with her brother in-law to increase her chances of getting pregnant [1]. She did not have access to treatment and in her culture she could not negotiate her status in her family and society outside motherhood. This is not a rare story rather the reality for many infertile women in developing countries. Infertility is not usually considered a developing world problem. Provision of infertility services and especially assisted reproduction are not on the resource allocation agenda. In the era of the HIV/AIDS epidemic, with malaria, tuberculosis and many other preventable or treatable diseases still claiming millions of lives, infertility can hardly make a case. At the same time, the number of private assisted reproduction technology (ART) clinics offering in vitro fertilization (IVF), intra-cytoplasmic sperm injection (ICSI) and other treatment is steadily rising [2]. But is infertility a real problem in the developing world?

Infertility affects around 10% of couples worldwide. This is a consensus estimate given the different definitions and methodologies that have been used in collecting international data about infertility [3]. Although lack of common definition, clear diagnostic categories and data collection methodologies create lack of precision, the data still show valid and remarkable differences between the developed and developing parts of the world. Developing regions score the highest infertility rates which in some countries go over 30% [4]. Variations among countries also exist between primary infertility (no pregnancy has ever being achieved) and secondary infertility (pregnancy has been achieved before). According to that distinction developed countries have a higher prevalence of primary infertility of roughly 3.6–14.3%

E. Vayena (✉)
Institute for Biomedical Ethics, University of Zurich, Zurich, Switzerland
e-mail: vayena@ethik.vzh.ch

A brief version of some of the ideas discussed here appear in Vayena E, et al. Assisted reproduction in developing countries; are we caring yet? (pending final acceptance of revised manuscript by Fertility and Sterility).

F. Simonstein (ed.), *Reprogen-ethics and the Future of Gender*, International Library of Ethics, Law, and the New Medicine 43, DOI 10.1007/978-90-481-2475-6_6,

while developing countries have a higher prevalence of secondary infertility [5, 6]. This phenomenon has been explained by the different etiologies in developed and developing countries.

The high prevalence of infertility in developing countries has been attributed to several causes such as environmental, dietary, exposure to toxic chemicals, smoking, alcohol and caffeine consumption [7–10]. However, the dominant cause of secondary infertility in developing countries is tubal factor due to pelvic infection from untreated sexually transmitted infections (STI), post-partum and post-abortion infections. Estimates for STI prevalence show that developing countries carry most of the global burden with sub-Saharan Africa being the region with the highest prevalence [11]. There is a strong correlation between high prevalence of STI and high prevalence of tubal factor infertility [3]. This correlation is critically important in the discussion about ART in developing countries for two reasons: first, ART (especially IVF) is the only solution to tubal factor infertility, second, STI which lead to tubal factor infertility are preventable diseases and therefore prevention of STI would mean prevention of most of secondary infertility. Male infertility contributes to at least half of all cases worldwide. For most of male infertility ART could offer solutions [12].

The Gender Dimension

The magnitude of infertility has dimensions beyond its prevalence and etiology. Bearing children is a core issue in people's lives; inability to achieve a pregnancy and live birth becomes a major problem, often a crisis, with psychological, financial and social consequences. Several studies showed that the infertility experience has significant negative effects on the individual woman and man as well as the couple and broader family [13–15]. Stress, depression, low-self esteem, guilt, marital problems, sexual problems have been documented as consequences of infertility. In that respect infertile people in the developed and developing world have something in common. However the differences are emerging again mainly for two reasons: (i) availability of infertility treatments and (ii) socio-cultural values surrounding procreation and infertility.

In general most developed countries provide infertility services including ART (although universal coverage of all ART treatments is not the case everywhere and low-income couples may be unable to afford ART). A recent analysis in western countries showed that 50% of infertile women seek medical help and they are usually of higher income and higher educational level [16]. In developing countries only a small fraction of the population is able to afford ART (provided that it is available in their country). Even though in most developing countries the prices are lower compared to those in western countries, treatment is still far from reach for many people living on medium income and certainly for those living around a dollar or less a day. The lack of choice makes the feeling of helplessness much more prominent.

What really exacerbates the social suffering of infertility in the developing world is the socio-cultural bias and discrimination against infertile people, especially women who are unable to have children [17]. In many developing countries having children goes beyond people's emotional needs, it is a matter of economic survival. Children are needed to work in farms, to help with daily tasks and support aging family members. Women are perceived as having the important role (and main responsibility) of producing children. Their status in society depends on their reproductive capacity and subsequently their well being and even life. In many such societies women carry the entire weight of the infertility problem, including the blame, irrespective of whether they carry the biological responsibility for infertility. There are significant variations among cultural responses to infertility but a review by van Balen and Gerrits showed that there are profoundly negative consequences for infertile women in many countries including Egypt, Nigeria, Mozambique, the Gambia, Zanzibar, Laos, and Bangladesh [18]. A study from rural Tanzania found that infertility was a cause of intimate partner violence against women [19]. Regions in the Indian sub-continent also present similar patterns [20]. In general, the consequences that infertile women experience include exclusion from social events, isolation, stigmatization and public humiliation, divorce, and domestic violence. Undoubtedly infertility in these regions is primarily a social problem and a gender issue and secondarily a medical condition.

A fertility doctor who worked for several years in Africa was quoted in the Nature article about Betty Chishava saying that "in Africa if you are a woman and you do not have children, you worth less than a dog" and argued for the urgent need to improve access to ART [1]. Whether treatment of infertility is the way to alleviate that kind of suffering can be contested. Two arguments can be made: the first is that provision of treatment alone risks to simply reinforce the stereotypical role of motherhood as a necessity for a woman's social status. ART can then be viewed as a quick fix of basic social problems that deem different solutions. Women's empowerment through education, increase of gender equality and other social changes would make a deeper and more sustainable contribution to the problem of infertility. Such changes require political will and a broader sociopolitical and cultural process, and they constitute the long-term approach. The second argument is that women who lack access to treatment are unable to exercise their reproductive autonomy. The argument is similar to the one used for access to contraception. In both cases women need access to services that enable reproductive choice [21]. However while reproductive autonomy is widely used in the discussion on access to reproductive health services, what remains unclear is how the concept of reproductive autonomy is actually constructed in non-western societies. For example how much is a woman's experience of reproductive autonomy shaped by the pronatalist society she lives in? If reproductive autonomy is not a concept pertinent to medical practice and societal thinking would ART ever constitute a real choice?

ART Clinics in Developing Countries: To Be or Not to Be?

Infertility clinics offering assisted reproduction technologies are available in many parts of the developing world. Several countries in Africa including sub-Saharan Africa, most South-American countries and India have numerous ART clinics with some of them performing high number of treatments and reporting success rates comparable to those in developed countries [2]. While it may still be odd that high technology infertility clinics are flourishing in countries which still lack the capacity to successfully deal with issues such as maternal and neonatal mortality or basic child vaccination programs, the reality is that ART clinics are most likely there to stay and in fact the number of them is on the rise. In addition some of these centers have already embarked on attracting international clients from developed countries by offering treatment at low prices. Western couples seeking ART treatment in India, or "mail order" egg-donation (with donors residing in developing countries) has brought international attention to the fact that the growth of the infertility industry in the developing world may have global implications [22, 23].

While all aspects of ART have been popular and highly debated topics in the western media and in the medical, legal, gender and bioethics literature, the issue of assisted reproductive technologies in developing countries and the special challenges it poses has received less attention. The fundamental issue that was partially examined is whether or not ART should be available and perhaps publicly funded and if it is, what does this entail. The predominant arguments have taken a negative take on the issue of provision. The most commonly used arguments are population growth, health priorities and costs.

The fact that population growth still remains a global issue has helped create the conviction that the problem of infertility cannot be given a high priority on the health agenda. The general perception is that countries with high fertility rates, as many developing countries have, do not have a pressing need for provision of infertility services. As I showed earlier this is false. Paradoxically, although with a sufficient explanation, countries with the highest fertility rates also have the highest infertility rates [3]. A better look at the world population trends shows that in general the world fertility rate has been reduced by half in the last fifty years but life expectancy has been rising, making the ageing population a stronger contributor to the population growth [24]. However to date, from a population point of view, controlling fertility through successful family planning programs has been the focus of national policies, aid organizations and international donors. This combined with the prohibitive costs that ART would incur to a health system have so far made a strong case against ART in developing countries. Where infertility has been acknowledged as a public health issue the response has focused on prevention (through prevention of STI post-partum and post-abortion infections) rather than treatment.

Other key points against ART provision have included: the pressing need to combat communicable diseases, decreasing maternal and child mortality, dealing with HIV, malaria etc.; the possible exploitation of infertile couples by offering unrealistic hopes given the limited success rates of ART; the physicians' limited training and the lack of appropriate infrastructure; the inability to deal with ART complications such as ovarian hypesrtimulation syndrome, multiple gestation, premature babies; reinforcement of stereotypical roles by pursuing motherhood at all costs; and finally ethical issues that ART raises and could challenge local religious and cultural norms [12, 18, 24–26]. Most of these arguments have been followed by a legitimate and well justified message: prevention of infertility through prevention of STI is preferable to provision of assisted reproduction. Given that most of the developing world infertility can be attributed to preventable causes this has been the widely favorite policy [27].

Recently some of these premises against ART in developing countries have been challenged. In 2001 the World Health Organization convened a meeting to discuss developments on ART but with a focus on their implications for developing countries. The meeting, which raised the issue for first time at that level and the published report did not answer the question of whether ART should become available in developing countries- that would have been UN unrealistic expectation. It did however succeed in attracting attention to the various facets of the ART in developing countries by providing a forum where regional views could be heard and interdisciplinary experts could deliberate recommendations. Such recommendations attempted to go beyond a black or white approach to ART in developing countries and called for more innovating approaches in resolving the problem such as development of low-cost ART, seeking alternative funding mechanisms where public funds are not available [28].

Building on some of these recommendations, the European Society for Human Reproduction and Embryology (ESHRE) established a task force on infertility and developing countries dedicated to document the problem and to explore ways and ART innovations that can be useful in the developing world [29]. One such initiative is the development and validation of low-cost ART treatments. Scientists have argued that it is possible to create ART protocols that are far less costly than the standard ones without substantially compromising the success of the treatment. While such work is still under development experts appear optimistic [1, 24]. In 2007 a scientific society was established under the name International Society for Mild Approaches in Assisted Reproduction (ISMAAR) with the intention to encourage the medical community to depart from the aggressive hormonal protocols that are commonly used. ISMAAR (by now a registered charity in the UK) argues that the costs of such ART protocols, the problem of multiple births and concerns about long-term health risks for the women undergo treatment can be resolved with simplified approaches. The society has also taken the issue of ART in developing countries and advocates "for IVF to be a safer, softer and affordable treatment globally" [30].

These developments indicate a shift on the approach of ART in developing countries in the last decade. In the 1990s ART met much more skepticism and ART in developing countries received a clearly negative answer [25, 31]. Today the response seems to be less rigid [24, 28]. The focus has moved on developing alternative types of ART which will be more appropriate for low-income setting. The idea of transferring ART as it is to the developing world is been put aside for the better suggestion of adapting ART so it can be a realistic option. Momentum is building on giving more justification to appropriate ART treatment in the developing world.

This apparent shift has not occurred in a vacuum rather within scientific, social and political changes. It is estimated that three million babies have already been born worldwide with the use of some form of ART, making ART a standard treatment of infertility [32]. Advances in ART have increased success rates and decreased complications related to ART such multiple pregnancies. Medical technology transfer has been accelerated through globalization. (High technology goods, for example mobile phones, have become widely available in the developing world). Several developing countries have made significant progress in improving health indicators and especially reproductive health indicators. For example Bolivia, Yunan in China, Jamaica and Egypt have reduced maternal mortality by half in the course of the last ten years [33]. Some developing countries are experiencing epidemiological transitions that allow them to reassess the priorities on their health agenda.

Furthermore, it is important to acknowledge that in the last two decades infertility consumer groups have emerged worldwide including in developing countries. Currently the international umbrella organization for infertility consumers called International Consumer Support for Infertility (iCSi) lists several national groups including at least four African groups such as in Nigeria, Kenya Uganda and Zimbabwe [34]. These groups have become an established and outspoken force on infertility issues. Their goals are to increase public awareness about infertility prevention and management and to push for provision of infertility services including ART in their countries. They oppose to the concept of ART as *luxury item* that is only within the reach of a few. They are active negotiators with health policy makers, scientists and the pharmaceutical industry [35].

The Future

The debate about ART in developing countries has reached a turning point. It moved away from the polarization of being *for* or *against* ART in developing countries to a more constructive phase of exploring new possibilities and innovative ways of making ART available in a manner that suits the needs and the particular situations of the developing world. Furthermore, it seems that there is more acknowledgement of the fact that individual health needs of impoverished people still have a place next to their public health needs. It seems that the issue will continue to move forward. I suggest three themes which construct a solid springboard for future steps in the discussion about ART in developing countries: provision of ART within the broader context of reproductive health and reproductive rights; ART within the heterogeneity of the developing world; and equity of access.

ART in Context

Discussions about ART in developing countries have focused on the legitimacy or not of ART provision in low-income settings without, however, successfully linking it with the need to improve reproductive health as a whole. The definition of reproductive health that was adopted by the United Nations International Conference on Population and Development (ICPD) states: "Reproductive health is a state of complete physical, mental and social well-being and not merely the absence of disease or infirmity, in all matters relating to the reproductive system and to its functions and processes. Reproductive health therefore implies that people are able to have a satisfying and safe sex life and that they have *the capability to reproduce and the freedom to decide if, when and how often to do so. Implicit in this last condition are the right of men and women to be informed and to have access to safe, effective, affordable and acceptable methods of family planning of their choice, as well as other methods of their choice for regulation of fertility which are not against the law, and the right of access to appropriate health-care services that will enable women to go safely through pregnancy and childbirth and provide couples with the best chance of having a healthy infant...*" [36]. According to this internationally accepted definition infertility is an important component of reproductive health. However, for most national programs reproductive health came to mean family planning programs, with access to infertility services been more on paper but hardly ever in action.

Any discussion about ART provision has a natural place in the context of reproductive health but this natural link has not been adequately argued. Perhaps a reason which contributed to this is the limited hard data about infertility's impact on reproductive health. For example, disability adjusted life years (DALYs), a health measure used for burden of disease calculations, cost utility analysis and health priority setting involves mortality and morbidity data. These measures have little relevance to infertility management because lives are not saved and disability is difficult to define and measure [37]. Better data and a more accurate portrayal of the burden of infertility (including the psycho-social suffering) could create a different framework for positioning infertility in the reproductive health agenda and for discussing ART in the developing world [38].

The public health approach for most diseases is through programs of prevention and treatment; infertility need not be argued as something different. Given the links between infertility and other reproductive health problems (STI, post-partum infections, and post abortion infections) advances in other areas of reproductive health would have an impact on infertility. It has been argued that countries that have succeeded in reducing post partum and post abortion infections are expected to see a decrease in infertility rates [39]. ART need to be seen and argued as a part of a more comprehensive reproductive health program that also includes prevention as well as appropriate diagnostic services. Prevention programs can reduce the prevalence of infertility, but prevention is an investment for the future and cannot alleviate the problem of those who are already suffering from it. Therefore prevention and treatment programs need to be developed in tandem. Arguing ART outside of that scheme would not hold sway, even if ART costs are reduced.

A renewed strategy and commitment to reproductive health was approved by the World Health Assembly in 2004, with the adoption of the first global reproductive health strategy [40]. Infertility is on the reproductive health agenda and although there is no reference to ART, this may present an opportune moment to frame the discussion on ART within the context of the renewed interest for improving global reproductive health.

Reproductive health as many other aspects of health have been increasingly discussed within the frame of human rights. Links between reproductive health and human rights have been extensively debated [41]. In relation to infertility two rights appear relevant: the right to marry and found a family as well as the right to benefit from the advancement of science and technology [42, 43]. The International Federation of Obstetricians and Gynaecologists (FIGO) specified the latter right further by stating that "women and men have the right to the highest available standard of health care for all aspects of their sexual and reproductive health" [44]. It was claimed that this can be taken further to establish a limited positive right to assisted reproduction with the implication that a fair and just society should make ART available to its citizens [45]. However for this to occur the society in question would need adequate resources. The issue of how the human rights approach could be used in the discussion of ART has not been adequately explored. Could the discussion become more constructive in the frame of reproductive rights? Is access to infertility treatment a reproductive right? Does this right clash with other human rights in settings with scarce health resources? Answers to some of these questions need urgent attention.

Heterogeneity of Developing Countries – One Size Does not Fit All

Developing countries are different. In international discussions and certainly in the discussion about ART they are often perceived as one homogenous part of world. Developing countries are heterogeneous in many ways such as culture, religious and moral values, disease patterns, political schemes and definitely on the pace of development [2]. Each of those factors impacts on the necessity to provide ART, on its potential availability as well as on its utilization by those in need. The differences are too many and too complex to accommodate generalizations. In the case of ART such generalizations have not been useful.

Differences are reflected well in the religious approaches. Religious views on ART are distinctly different on profound ethical issue in relation to the status of the embryo, gamete donation, surrogacy etc. The Catholic Church has prohibited ART, Judaism has been favorable, Islamist fatwas have endorsed some aspects of ART, Hinduism has not addressed the issue but seem favorable, Protestantism has accepted most of ART [46]. Interestingly, Israel is one of the few countries in the world (along with some Scandinavian countries) that have optimal utilization of ART with the highest ratio of assisted reproductive technology clinics to population [37, 46]. On the other hand Latin American countries have been heavily influenced by Catholicism in their policies and laws related to human reproduction. It cannot be a mere coincidence that Costa Rica's Supreme Court outlawed ART because it judged it unethical [47]. However, ART is available in many other Latin American countries and as data from the Latin-American register show the number of ART treatment has steadily increased over time [48].

Another interesting cultural variation that may impact utilization of ART is cultural perceptions of adoption. Studies have showed that some cultures have deeply-rooted resistance to the idea of raising an adopted child. For example in Egypt adoption restrictions are culturally constructed on fears about the immorality of "illegitimate" children that need a foster family, lack of kinship and affinity, social stigma that the couple is unable to produce a real child etc. [12]. Couples in Egypt would rather resort to ART than to adoption. In India social pressure to keep infertility secret also seems to lead infertile couple to ART clinics. There they might be able to overcome their problem maintaining their privacy while adopting would be a visible admission of infertility [49]. In contrary, for some African countries adoption is a realistic and culturally viable option. Especially fostering arrangements through relatives are viewed as a partial solution to infertility [50].

Finally, developing countries do not have the same pace of development and therefore they vary on their capacity to carry on technologically demanding medicine. The first IVF baby in India was born in 1986 within a government research project of the National Institute for Research and Reproduction [22]. The first IVF baby of Nigeria in 1989 at the Lagos University Teaching Hospital. (The world's first IVF baby was born in 1978 in England) [28]. It is imperative to discuss ART in the developing world keeping in mind the differences among countries and the powerful connections between socio-cultural beliefs, policy and provision and usability of services. In an increasingly globalised world it might be tempting to act as if one size fit all but at least in the case of ART there is enough evidence pointing to the fallacy of such an approach.

The Equity Issue

While there are clearly many factors that influence availability and utilization of ART services, the decisive factor is cost. Assisted reproduction comes in a high price and that is a global truth. As a result, in most countries, ART is not covered by public health insurance (or even private insurance) and in developing countries assisted reproduction is usually available only outside government financed facilities [3]. ART is far beyond the means of many infertile people and mainly available to the well-off. A recent study on the international health economics of ART (IVF and ICSI) showed that global provision of ART corresponds only to a fraction of the projected utilization even in developed countries, with less than 20% of those needing IVF/ICSI actually utilizing it [37]. Developing countries face the biggest problem with only about 1% of the projected need. The study also showed that IVF costs were 50% above the gross national income per capita of most developing countries studied, specifically of China, India, Indonesia, Iran, Jordan, Lebanon, Malaysia, and Pakistan. As long as ART costs remain at such levels, ART will continue being available only to the financially well-off. The poor will not have access to ART and even those with medium income will have to face catastrophic financial consequences if they would seek treatment. Existing disparities in access will continue existing and growing.

Fortunately there is much discussion today about the possibilities of reducing ART costs in general and of developing approaches that can be used in poor settings [51]. In fact this is the most notable shift in the way that ART in developing countries is debated. The new approach suggests the use of less expensive hormonal protocols, different types of drugs, different kind of equipment [24, 30, 52]. The goal of such innovative approaches is to reduce the costs and infrastructural requirements of ART so they can be performed in poor settings. If this possibility materialize it might be more likely for health systems to fund ART, for public-private partnership to develop and maybe for more individuals to purchase them.

This is a positive development, however the question of how affordable could ART become still creates some skepticism. For example in Nigeria one IVF cycle costs $2000–$2700. The minimum wage is $52–$60 per month [28]. Is this a gap that can be bridged? The skepticism should by no means hurdle the new developments but should contribute into making equity a priority in the debate. There is a moral responsibility to increase access to health services. Increase of access is a desirable goal but whose access increases is a key question. Although not well documented (except from Latin American countries which contribute to the Latin American Register) the number of clinics that provide ART increases and so do the numbers of treatments. Could this increase in absolute numbers indicate any decrease of disparities? It is impossible to answer the question without data and the issue has to date not been examined but an educated guess would give a negative answer. Lessons from other areas of health would advice caution on how to interpret national averages. For example, increase in contraceptive use at national averages did not translate into equal increase for all wealth quintiles and in fact showed that the worse-off had made no progress in contraceptive use [53]. If the goal is to bridge the gap in disparities, it is critical to ensure that potential increase of access should happen across the board and should not only facilitate access for those that anyway had it.

An issue closely linked to costs is that of quality. The innovative approaches that are currently proposed are opting to maintain high quality treatment and good success rates. Minimizing risks and maximizing benefits should remain the guiding principle. Innovative and affordable ART need to ensure a favorable risk/benefit ratio for the infertile couple. The less costly approaches should not create a *double-standard* whereby the poor get poor quality and the rich the best quality. This would be unjust and would increase equity gaps between poor and rich. Many developing counties have limitations in quality and delivery of care and such limitations need to be taken into account in the development of the affordable ART. Equity of access is critically important but equity of access to good quality of care is what will make the difference.

Conclusions

The debate about assisted reproduction in developing countries has entered a new more constructive phase. It moved away from the narrow notion of ART as a *luxury item* not suitable for impoverished people in the developing world. This is a result of broader scientific, social and political changes. Initiatives to make ART more affordable and technically appropriate for poor resource settings are gaining momentum. Whether they will be successful remains to be seen. I have argued that to increase the chances of success future steps should ensure that the justification of ART for the developing world should be framed within the broader commitment to improve reproductive health; should acknowledge the socio-cultural differences of developing countries as well as their different phases in development; and finally should ensure that equitable access to high quality of ART services is the driving force behind any further developments.

References

1. Pilcher H (2006) Fertility on a shoestring. Nature 442:975–977
2. Vayena E et al. (2002) Assisted reproductive technology in developing countries: why should we care? Fertil Steril 78:13–5
3. Nachtigall RD (2006) International disparities in access to infertility services. Fertil Steril 85:871–5
4. Rutstein S (2004) Infecundity, Infertility, and Childlessness in Developing Countries. DHS Comparative Reports No. 9. Calverton, Maryland, USA: ORC Marco and the World Health Organization
5. Schmidt L, Münster K (1995) Infertility, involuntary infecundity, and the seeking of medical advice in industrialized countries 1970–1992: A review of concepts, measurements and results. Hum Reprod 10:1407–18
6. Lunenfeld B, Van Steirteghem A (2004) Bertarelli Foundation. Infertility in the third millennium: implications for the individual, family and society: condensed meeting report from the Bertarelli Foundation's second global conference. Hum Reprod Update 10:317–26
7. Hamberger L, Janson PO (1997) Global importance of infertility and its treatment: role of fertility technologies. Int J Gynaecol Obstet 58:149–58
8. Ibeh IN et al. (1994) Dietary exposure to aflatoxin in human male infertility in Benin City, Nigeria. Int J Fertil Menopausal Stud 39:208–14
9. Mattison DR et al. (1990) Reproductive toxicity: male and female reproductive systems as targets for chemical injury. Med Clin North Am 74:391–411
10. Turesky RJ et al. (2003) The effects of coffee on enzymes involved in metabolism of the dietary carcinogen 2-Amino-1-Methyl-6-Phenylimidazo[4,5-B]Pyridine in rats. Chem Biol Interact 145:251–65
11. World Health Organization (2001) Global prevalence and incidence of selected curable sexually transmitted diseases: overview and estimates. Geneva: WHO
12. Inhorn M (2002) Global infertility and the globalization of new reproductive technologies: illustrations from Egypt. Soc Sci Med 56:1837–51
13. Greil AL (1997) Infertility and psychological distress: a critical review of the literature. Soc Stud Med 45:1679–1704
14. Hardy E, Makuch MY (2002) Gender, infertility and ART. In: Vayena E, Rowe P, Griffin D (eds.) Current Practices and Controversies in Assisted Reproduction- Report of a World Health Organization Meeting. Geneva: WHO
15. Van Balen F, Trimbos-Kemper TCM (1993) Long-term infertile couples, a study of their well-being. J Psychosom Obst Gynaecol 14:S53–S60
16. ESHRE Capri Workshop Group (2001) Social determinants of Human Reproduction. Hum Reprod 16:1518–26
17. Daar A, Merali Z (2002) Infertility and social suffering: the case of ART in developing countries. In: Vayena E, Rowe P, Griffin D (eds.) Current Practices and Controversies in Assisted Reproduction Report of World Health Organization Meeting. Geneva: WHO

18. Van Balen F, Gerrits T (2001) Quality of infertility care in poor-resource areas and the introduction of new reproductive technologies. Hum Reprod 16:215–9
19. McCloskey LA et al. (2005) Gender inequality and intimate partner violence among women in Moshi, Tanzania. Int Fam Plan Perspect 31:124–30
20. Widge A (2002) Sociocultural attitudes towards infertility and assisted reproduction in India. In: Vayena E, Rowe P, Griffin D (eds.) Current Practices and Controversies in Assisted Reproduction- Report of a World Health Organization Meeting. Geneva, WHO
21. Germain A (2004) Reproductive health and human rights. Lancet 363:65–6.
22. Mukherjee M, Nadimipally S (2006) Assisted reproductive technologies in India. Development 49:128–34
23. Heng BC (2006) The advent of international 'mail-order' egg donation. Br J Obstet Gynaecol 113:1225–7
24. Ombelet W, Campo R (2007) Affordable IVF for developing countries. Reprod Biomed Online 15:257–65
25. Okonofua FE (1996) The case against new reproductive technologies in developing countries. Br J Obstet Gynaecol 103:957–62
26. Macklin RB (1995) Reproductive technologies in developing countries. Bioethics 9:276–82
27. Reproductive Health Outlook (2004) Infertility http://www.rho.org/html/files/RHO_infert_12-19-04.pdf#search="infertility" Accessed January 12 2008
28. Vayena E, Rowe P, Griffin D (eds.) (2002) Current Practices and Controversies in Assisted Reproduction- Report of a World Health Organization meeting. Geneva: WHO
29. European Society for Human Reproduction (ESHRE). Task force for infertility in developing countries. http://www.eshre.com/emc.asp?pageId=1040 Accessed 20 January 2008
30. Nargund G, Frydman R (2007) Towards a more physiological approach to IVF. Reprod Biomed Online 14(5):550–2
31. Heitman E (1995) Infertility as a public health problem: why assisted reproductive technologies are not the answer. Stanford Law Pol Rev 6:89–102
32. Personal communication with Dr Karl Nygren, chair of the International Committee on Monitoring ART (ICMART) based on data from International Committee for Monitoring Assisted Reproductive Technology (ICMART) et al. (2006) World collaborative report on in vitro fertilization, 2000. Fertil Steril 85:1586–622
33. Liljestrand J, Pathmanathan I (2004) Reducing maternal mortality: can we derive policy guidance from developing country experiences? J Public Health Policy 25:299–314
34. International Consumer Support for Infertility (2008) About international consumer support for infertility. http://www.icsi.ws/about_icsi Accessed January 30 2008
35. Dill S (2002) Consumer perspectives. In: Vayena E, Rowe P, Griffin D (eds.) Current Practices and Controversies in Assisted Reproduction Report of World Health Organization Meeting. Geneva: WHO
36. United Nations Population and Development (1995) Programme of Action adopted at the International Conference on Population and Development: Cairo: 5–13 September 1994. New York: Department of Economic and Social Information and Policy Analysis, United Nations
37. Collins J (2002) An international survey of the health economics of IVF and ICSI. Human Reproduction Update 8:265–77
Oknofua FE (2003) New reproductive technologies and infertility treatment in Africa. Afr J Reprod Health 7:7–11
38. Penning G (2007) Ethical issues in infertility in developing countries. http://www.eshre.com/file.asp?filetype=doc/04/014/001/001/7_s_arusha_infertility_2007.pdf Accessed 17 January 2008
39. Aboulghar M (2005) The importance of fertility treatment in the developing world. Br J Obstet Gynaecol 112: 1174–6
40. World Health Organization (2004) Reproductive health strategy to accelerate progress towards the attainment of international development goals and targets. http://www.who.int/reproductive-health/publications/strategy.pdf Accessed 12 January 2008

41. Cook R, Dickens B, Fathalla M (2003) Reproductive health and human rights. New York: Oxford University Press
42. United Nations (1976) International Covenant on Economic, Social and Cultural Rights. http://www.unhchr.ch/html/menu3/b/a_cescr.htm Accessed January 30 2008
43. United Nations (1948) Universal Declaration of Human Rights. http://www.un.org/Overview/rights.html Accessed January 30 2008
44. FIGO Committee for the Study of Ethical Aspects of Human Reproduction and Women's Health (2006) Ethical Issues in obstetrics and gynecology. http://www.figo.org/docs/Ethics%20Guidelines%20-%20English%20version%202006%20-2009.pdf Accessed 20 January 2008
45. Chan CCW, Ho P-C (2006) Infertility, assisted reproduction and rights. Best Pract Res Clin Obstet Gynecol 20:369–80
46. Chavkin W (2006) The old meets the new: Religion and assisted reproductive technologies. Development 49:78–83
47. Luna F (2002) Assisted reproductive technology in Latin America: some ethical and socio-cultural issues. In: Vayena E, Rowe P, Griffin D (eds.) Current Practices and Controversies in Assisted Reproduction- Report of a World Health Organization Meeting. Geneva: WHO
48. Zegers-Hochschild F (2002) The Latin American registry of assisted reproduction. In: Vayena E, Rowe P, Griffin D (eds.) Current Practices and Controversies in Assisted Reproduction-Report of a World Health Organization Meeting. Geneva: WHO
49. Bharadwaj A (2003) Why adoption is not an option in India: the visibility of infertility, the secrecy of donor insemination, and other cultural complexities. Soc Sci Med 56:1867–80
50. Gerrits T (2000) Studying infertility and STD's among the Macua in Montepuez, Mozambique: A multidisciplinary challenge. In: Van Balen F, Gerrits T, Inhorn MC (eds.) Proceedings of the conference on social science research on childlessness in a global perspective 8–11 November 1999, Amsterdam, The Netherlands. Amsterdam: SCO-Kohnstamm Instituut, University of Amsterdam
51. Pennings G, Ombelet W (2007) Coming soon to your clinic: patient-friendly ART. Human Reproduction 22:2075–9
52. Hovatta O, Cooke I (2006) Cost-effective approaches to in vitro fertilization: means to improve access. Int J Gynaecol Obstet 94:287–91
53. Gakidou E, Vayena E (2007) Use of Modern Contraception by the Poor is Falling Behind. Plos Medicine 4:381–9

Chapter 6
Ethical and Socio-Cultural Aspects that Influence ART in Latin America

Ester Polak de Fried

Introduction

The medical advances of the Assisted Reproductive Technologies (ART) divulgation in the last decades have awoken an enormous interest in the population in general because they are closely linked to human's procreation and sexuality and their future development. Their have also put in evidence the ever-going universal debate about the beginning of life.

ART set up continuous challenges which impact on social norms, moral and ethical standards and legal systems. They also generate great controversies at individual, social and political-legal level. All this, added to the media exploitation of the topic, has transformed the issue of how ART can have a greater and an even more important influence on worldwide topics.

The opposite situation of how the world in general, Latin-American regions and countries in particular and their different policies can impact on ART are not as well known and, hence, less discussed.

The world is not homogeneous; demographic, social, political, media, religious and economic aspects are different and each of them has a direct action in ART policies and politics. Therefore, we can say that the world is heterogeneous according to its population in general, its population density in particular, gross domestic product, undernourished population, infant mortality, birth rate and religion, just to name a few [1].

All the parameters mentioned above divided the countries in developed and developing countries. Developed countries have developed economies, domination of tertiary, domination of quaternary sectors of industry, high human development index and high gross domestic product, while developing countries have relatively low standard of living, undeveloped industrial base, widespread poverty, moderate to low human development index and low gross domestic product [2].

E. Polak de Fried (✉)
CER Medical Institute, Department of Reproductive Medicine, School of Medicine, University of Buenos Aires, Buenos Aires, Argentina
e-mail: ester_polak@cermed.com

F. Simonstein (ed.), *Reprogen-ethics and the Future of Gender*, International Library of Ethics, Law, and the New Medicine 43, DOI 10.1007/978-90-481-2475-6_7,

The main differences can be seen comparing the gross domestic product and the human development index. In a population pyramid for a developed country the population from 0 to 4 years is very similar to the one from 60 to 70 years. There is low fertility rate, low infant mortality and high life expectancy. In a population pyramid for a developing country the infant population is very wide but the pyramid grows narrow geometrically. There is high fertility rate, high infant and maternal mortality and low life expectancy. The number of people who gets to live longer is extremely reduced [3].

In the next items, we will deal with how these differences will influence the development, or not, of diseases that produces infertility. How demographic changes affect different economical policies that, in time, determine, or not, standards to incentive fertility rates and indirectly the ART in variety of ways. A direct reference towards the impact of market crisis on family planning will be made. Finally, the present state of the different reproductive medicine aspects in Latin America will be developed.

Continents and Countries

It is already established that the world is not uniform, not even the continents. A number of countries, when referring to the United States, call it "America", when in fact, the American continent includes several countries that are part of North America and Latin America. Something similar happens in Europe. Nowadays there are two Europes; one runs from the Mediterranean to Central and Eastern Europe and the other stretches from Scandinavia to France.

According to some demographic aspects the United States and Europe are usually compared. In 2002 the United States had a great advantage over Europe according to the Fertility Rate and there were studies about the impact this would have on the economic and political future. Suddenly, five years later, in 2007, all the information seemed to show that, in part, the Old World looks younger because some important countries like France presented the same index as United States [4, 5]. Nowadays there is a marked demographic crisis in the Fertility Rate upon Europe as well as the United States and Canada [6]. Moreover, the two Europes determine the fertility rate of the region.

Some countries took measures about this situation. For example, Germany promotes generous grants and longer working licenses to increase parenthood. This was exclusively address to the middle class and professionals [7]. Spain's President declared that "Each family living legally in Spain will receive 2,500 euros from today for every new child born in our country" [8]. In June 2007, women in the Russian city of Ulyanovsk stand to win a new car; but only if they gave birth on exactly June 12th: "Russian National Day". The city authorities have officially declared September 12th, 2007, the "Day for Family Communication" [9].

When Germany refers to the middle class and professionals, and Spain refers to those who live legally in the country, there is a direct relation with the migratory and illegal movements that affect these countries in different ways. It also reveals a clear politic to benefit the population growth of legal citizens.

These policies will be probably adopted by a group of people that had postponed their parenthood and even those who had chosen not to have kids, being the average age of this group of people from 35 to 45.This could be probably reflected in a major ART demand. According to the average age of this group, an increment of bioethical controversies can also arise.

Although Italy is going throughout a tremendous crisis to renew the population because the fertility rate is very low, in the year 2004 a very restrictive law towards ART was created. Many ART procedures are not allowed and there are several concerning items, such as: they only allow to fertilize three oocytes, they should transfer all the pre-embryos achieved, embryo cryopreservation is not allowed, they consider a person to exist with the only contact of sperm and oocyte [10]. This has created some ethical dilemmas in reproduction: infertile couples have lower chances to get pregnant, there is an encouragement of "procreative tourism", there is more risk of hypermultiple pregnancies and there is an important controversy about the status of pre-embryo, embryo, fetus and born person.

Francoise Shenfield [11] makes a very interesting analysis about these aspects bearing in mind the coverage of the public health system and the fertility treatments. The imbalance between supply and demand faces a problem of justice, as to whom is selected as deserving of subsidize treatment, and a problem of policy, as to who may decide on this access to treatment. She argues there is little distance between policy and politics referring to the creation of the National Health Services in the UK, based on the ideal of equal access to treatment for all citizens in need.

In the year 2006, a clear expectancy of population growth was installed in the United States with the aim of solving more economic problems and increasing population and consumption. In the starting of the second trimester of 2007 the great break of the financial market came up and drastically shook the United States' economy and impacted on the rest of the world [5]. Would this evolution have impacted on that young population, in a fertile age, that might have invested in mortgages to program a family? If this is correct, the effect will be shown in a decrease of the Fertility Rate. We will know this in a couple of years.

In Asia, Japan marks a difference for being a developed country. Its Human Development Index differs from the rest. Recently the concern about how to struggle with a shrinking population arose and worries about a population explosion have been replaced by fears of decline. It is very clear that the young people are disappearing and the forecast of the population pyramid is practically turned upside down. Many economist advisors stated that if a country wants to keep its population up, it should promote In Vitro Fertilization (IVF).

More and more there are shrines where many of Japan's older people go to pray not only for the long life that they are increasingly coming to expect, but also for a quick and painless death. There is an increase in the number of these shrines since they are the most popular and the most visited ones [5].

Worldwide, more than 80 million couples suffer from infertility; the majority of them are residents of developing countries and Africa. One out of three couples is infertile in Sub-Saharan Africa, 9% in Gambia, 21.2% in Ethiopia and 20–35% in Nigeria. A study by the WHO indicates that most of infertility cases in Africa are caused by tubal occlusion secondary to infections, most frequently caused by

Chlamydia, gonorrhea and/or tuberculosis. Laparoscopy revealed a prevalence of bilateral tubal occlusion in 35–65% of infertility cases in Nigeria. In Durban a tubal factor was identifiable in 77% of the infertile population [12].

Latin America

As it was said before, a part of the American continent is developing. Latin America is a developing region very different from the rest of the continent. Latin America is usually considered a homogenous region. Actually, the countries of this region have cultural and sociological individualities which make them different. Although Latin America is not a total homogenous region since its population differs on its origins (Africans, Indigenes, Europeans), natural resources, sizes; they have in common an increase of the difference between social classes and the predominance of Catholic religion. As there are other priorities of the region considered to be solved, some politic speeches argue that ART are a good of luxury.

The economical, political and social reality in the Latin-American countries generates more people who suffer from infertility caused by the lack of access to appropriate primary health resources.Unfortunately, people of the poor classes with low resources have no or less access to good health services, which produce an increase of sexually transmitted diseases (STD) and reproductive tract infection (RTI). The lack or less access to contraception leads also to an increase of STD and RTI that produce infertility, and also leads to illegal abortions in deficient conditions usually followed by the death of these young women or, in a better framework, followed by psychological and tubal damage.

Despite the fact that Catholicism dominates the countries in the region, strangely enough, the number of illegal abortions is high (Argentina far exceeds 500 thousand per year). A Brazilian study showed that 42% of women who consulted for infertility had tubal obstruction because of reproductive tract infection (RTI) [13]. Working conditions like the exposure to toxics also leads to fertility problems. It was reported that indigenous people from Guatemala after the exposure to fertilizers had a high incidence of male factor problems.

The multiple factors of environment that could cause major adverse effects on human being was a clear incentive for Prof. Benyakar to invite professionals in different fields to participate in the Iberoamerican Network of Eco-Bioethics from UNESCO Chair, in which these topics are developed from the reproductive medical scenario [14]. In most of these countries, both the public and the private health systems seem to forget about the problematic of infertile couples. The population with fertility problems is part of the society as well.

It is very hard to understand the unequal access of the Latin-American infertile couples not only to fertility treatments but also to an adequate sexual education and the implementation of contraception. Most of this situation depends on the different current governments and their relationship with the Catholic Church [15]. In Argentina, there is a close relationship between the Church and the State, but the National Constitution maintains religious freedom. The same holds good for Bolivia, Peru and Colombia. The National Constitutions of Chile, Ecuador and Mexico only state freedom of choice, and both Brazil and Uruguay propose, directly, a Secular State [16].

During the 90s, the influence of scientists' affairs consultant, Cardinal Ratzinger, in his speech at the Congregation for the Doctrine of the Faith, awakened in certain groups an evident aversion to the New Reproductive Technologies [17]. A persistent and aggressive debate prevented ART law regulation from being passed [18]. Curiously enough, the very same Ratzinger, the present Pope Benedict XVI, on February 1st 2008, using practically the same speech after two decades, attacked all kind of fertility techniques and claimed that: "They attack the very essences of human dignity". The rejection was evident in a speech that was addressed to the Congregation for the Doctrine of the Faith, in Rome.

Benedict XVI claimed that these fertility methods attack the very essence of human dignity. "When human beings, in the weakest and most defenseless state of their existence, are selected, killed, abandoned or used as biological material, how could anybody deny that they have already been treated as something and not somebody, thus questioning the very concept of human dignity?" [19]

It is evident that in countries like Argentina, these statements will once more raise controversial opinions aiming at stopping the level of development in an area that has enabled thousands of couples to have a family. Obviously, the society as a whole must be alert to avoid restrictive lines of though that limits human rights and freedom of choice [20]. Likewise, this also restricts an adequate family planning that is within reach of all social classes. Each year in Argentina 3.000 babies are born from children-mothers. Girls between 10 and 14 years old are many times victims of sexual abuses. A study indicates that the fertility rate at this age has duplicated in 40 years. Infant mortality is quadruple in mothers under 15 years old.

The main cause is the mum's immaturity that makes her not to be alert: she couldn't take care of herself; she could hardly take care of a baby. The drama of the children-mothers is not homogenous throughout the country: Buenos Aires and Santa Cruz states have values lower than 2%, while in Catamarca, Corrientes, Chaco, Formosa and Santa Fe the values are higher than 4%. Poor women are the main victims of this kind of maternity. Among analphabets, the adolescent pregnancy rate goes from 11% to 25% (it triplicates in women under 15 years old) [21].

In Latin America, not a single country has the use of ART governed by law; some countries are regulated by guidelines and others have no regulation at all. Countries in the rest of world are regulated by law, by guidelines and of course others have no regulation. More interesting is to look at the coverage or reimbursement of ART. While the countries of the rest of the world have some kind of coverage, the Latin-American countries have no coverage of any kind [10] (Figure 6.1).

Not only in Latin-America, but also in other regions, the records aren't faithful to the reality. That's why it's necessary to consult many sources to have a very clear idea about what happens to ART in the region. In comparison between the IFFS Surveillance 2007 [10] and the Latin-American Network of Assisted Reproduction (RED-LARA) [22] we found that 9 of the 57 countries which completed the questionnaire of the Surveillance are from Latin America: Argentina, Brazil, Chile,

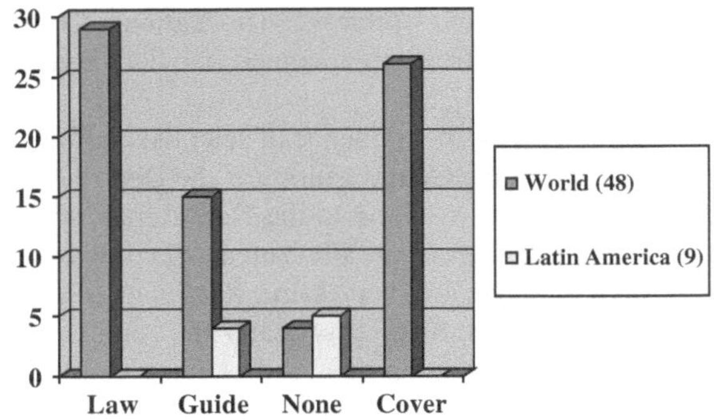

Fig. 6.1 Differences between Latin America and the Rest of the World regarding ART

Colombia, Ecuador, Mexico, Peru, Uruguay and Venezuela. From these 9 countries the approximate number of centers is 263. The RED-LARA receives information from different Latin-American centers, which wish to report voluntarily and take part in the Registry. The number of countries, which reported to the RED-LARA, is 10, to the previous countries mentioned Guatemala is added, and the number of centers that reported information is 128 (Table 6.1).

From the analysis effectuated, it is clear that a high number of existing centers in Latin America did not report information to the RED-LARA. The number of procedures effectuated in Latin America according with RED-LARA's 2007 report was 24,588. In a comparative analysis we could think that with the information from the Surveillance (which is not a registry since it doesn't request number of procedures), the total number of procedures effectuated in Latin America would be much more than 50 thousand per year.

Costa Rica is the only Latin-America country that has, by constitutional amendment, forbidden In Vitro Fertilization and this doesn't appear in the records. A petition was recently filed in the international ethic community claiming that that prohibition violates the Declaration of Human Rights. Until the presentation date of this chapter, the results have not been known.

A very curious situation is presented to the embryo cryopreservation. Embryo cryopreservation isn't forbidden in many countries. Nevertheless, a lot of centers from many countries don't implement it. Consequentially, this could increase hypermultiple pregnancies risk and perinatal morbimortality.

This is another example of Latin America's big receptivity to comply with the Church's designs. In 1993, in Argentina, the lawyer Ricardo Rabinovich promoted a precaution action to control everything relating to embryo cryopreservation. According to him, the Civil Code "where embryos are persons and whatever is done or not with them can not be decide by the parents, laboratories or doctors", charging all IVF Centers from cryopreserving persons without their authorization. He asked Justice to guarantee "the physical and spiritual integrity of these absolute incapables" referring to the cryopreserved embryos.

A decade later he was designated by a judge as "special tutor of cryopreserved embryos and pronucleated oocytes" [23]. This situation produced an important

Table 6.1 Differences between IFFS Surveillance and RED-LARA in number of centers and procedures

	Number of centers		Number of procedures	
Countries	IFFS Surveillance	RED-LARA	RED-LARA	Real (Aprox.)
Argentina	20–26 (23)	21	5,083	–
Brazil	105–200 (152)	55	10,804	–
Chile	8	7	1,201	–
Colombia	17–18 (17)	9	1,102	–
Ecuador	6	3	177	–
Mexico	40	22	3,797	–
Peru	4	2	1,047	–
Uruguay	4	2	429	–
Venezuela	5–12 (9)	6	918	–
Guatemala	–	1	30	–
Total	**263**	**128**	**24,588**	**50,000**

reaction of all IVF Centers and the prospective parents of those cryopreserved embryos, as well as of the Argentinean Society of Human Reproduction, hindering its development. The guardianship of the embryos is linked to the roman *curaventris*. The *curator ventris* was a citizen elected to look after the health, life and interests of the child in gestation and the mother [24].

Bioethicist Florencia Luna summarizes the above situation stating that "the embryo is sometimes more protected than women in Latin-American countries" [16].

Conclusions

Infertility affects some 168 million people worldwide and has no boundaries of race, standard of living or social status. Developing countries are the least likely to have access to an adequate sex education and, as a result, have high infertility rates for treatment. Many international organizations have issued their opinion on this subject [20, 25, 26]. *Prevention* and *education* remain the most important objectives considering the issue of infertility in developing countries. To be successful family planning, *prevention* of infertility as well as *diagnosis* and *treatment* of infertility should be incorporated in existing *health care centers* [12].

As responsible professionals we must bear in mind that the present generation has the responsibility of safeguarding the needs and interests of future generations. Education should be used to foster peace, justice, understanding, tolerance and equality for the benefit of present and future generations [27].

Bob Edwards, along with Patrick Steptoe, was responsible for the first in vitro birth in 1978. During a recent conference, Edwards cleverly said: "Biologists are not on the edge of destructive items such as the atom bomb. We are pro-life, although some recent ideas in biomedicine imply that it is time to learn a little more about ethics and moral philosophy. Remember what Jacques Monod wrote about our living in a universe which fails to listen to our music. We are essentially isolated and must devise the best medical treatments for members of our species, while ensuring the highest standards of ethics and moral philosophy" [28].

Either way, there are still questions unanswered: Why are the patients' rights in general and justice in particular so vulnerable throughout the world? How far the professionals are aware of this situation and take part in a committed way to preserve the rights of the patients? We must clarify that no society can be considered modern (open-minded and with changeable values) if there is no freedom of election, freedom of reproduction and medical responsibility.

References

1. Encarta (2005)
2. United Nations (2006) Development Program. Encarta. Acceded on line
3. Standard Grade Geography (2001) Encarta
4. The Economist (2002) August 24 http://www.economist.com/
5. The Economist (2007) July 26 http://www.economist.com/
6. Pan-American Health Organization (2007)
7. Merkel A (2006) First European Forum about Demographic Future. December 15th 2006, Brussels
8. Zapatero R, (2007) El País newspaper. March 13, http://www.elpais.com/articulo/espana/Gobierno/confirma/solo/cobraran/2500/euros/nacidos/julio/elpepuesp/20070713elpepunac_2/Tes Spain
9. Morozov (2007) La Nación newspaper. September 13, http://www.lanacion.com.ar/archivo/nota.asp?nota_id=943529&origen=acumulado&acumulado_id Argentina
10. Jones HW, Cohen J (2007) IFFS Surveillance 2007. Fertil Steril, Vol. 87, No. 4, Suppl. 1
11. Shenfield F (1997) Justice and access to fertility treatments. In: Shenfield F and Sureau C (eds) Ethical Dilemmas in Assisted Reproduction, 1st edn. London, UK
12. Ombelet W, Campo R (2007) Affordable IVF for developing countries. RBM Online, Vol. 15, No. 3:257–265
13. Fernandez MS, Bahamondes L (1996) "Incidência dos fatôres etiológicos de esterilidade conjugal nos hospitais universitários de Campinas", Revista Brasileira de Ginecologia e Obstetricia 18:29–36
14. Benyakar M (2005) Meeting of The Iberoamerican Eco- Bioethics Network for Education, Science and Technology. The UNESCO Chair in Bioethics 23th August 2005. School of Law, UBA. Buenos Aires, Argentina
15. Polak de Fried E (2006) Justice and socio-cultural issues regarding infertility and ART in Latin America. IFFS Newsletter, Autumn, 3–4
16. Luna F (2002) Assisted reproductive technology in Latin America: some ethical and sociocultural issues. In: Vayena E et al. (eds) Current practices and controversies in Assisted Reproduction. 1st edn. WHO, Geneva, Switzerland
17. Ratzinger J (1990) Congregation for the Doctrine of the Faith, 1990. The instruction on respect for human life in its origin and on the dignity of procreation, Ed. San Pablo
18. Polak de Fried E (1995) Assisted reproductive techniques. In: Bulletin of the Library of the Nation's Congress No 118. Pag. 149–158. Buenos Aires, Argentina
19. Ratzinger J (2008) Congregation for the Doctrine of the Faith, Rome. February 1, Permanent Link http://www.mdzol.com/mdz/nota/29036
20. United Nations (1948) Universal Declaration on Human Rights. New York. Art 16 and 25 http://www.un.org/Overview/rights.html
21. Argentinean National Health Ministry (2007)

22. Zegers-Hochschild F et al. (2007) Latin American Registry of Assisted Reproduction 2003–2004. Santiago, Chile
23. El Clarín Newspaper (2005) 26th August. www.clarin.com/diario/2005/07/26/sociedad/s-03101.htm
24. Digesto (530 b.C.) Compilation and codification of the legal masterpieces of the roman jury consultants. La Ley, Buenos Aires, Argentina
25. Council of Europe (1950) Convention for the protection of human rights and fundamental freedoms. Article 12. Rome, Italy
26. UNESCO (2005) Universal Declaration on Bioethics and Human Rights. Adopted by acclamation on 19 October 2005 by the 33rd session of the General Conference of UNESCO. Paris, France
27. Carmi A (2007) International Conference on Bioethics: Today in the mirror of future generations. Eilat, Israel
28. Edwards RG (2007) Reminiscences on learning about morals and ethics in biomedicine. Ethics, Science and Moral Philosophy of Assisted Human Reproduction, Vol. 14, Suppl. 1

Part III
Gendered Futures

Chapter 7
Women, Work, and Children: Is There a Solution?

Daniel Callahan

Introduction

For over five decades, there have been few problems more vexing for women than how to balance work and family life. To be sure, women have always worked, usually out of necessity. But the rapid increase in higher education for women after World War II brought a massive change in possibilities, opening the way to work by choice and to seek the same professional goals as men. An important part of that story is for women who want to work to determine whether and when to have children, and how many of them to have. The number of books and articles on the subject must run into the thousands, most of them written by women and for women. There is also a large technical literature coming out of the fields of demography, sociology, and women's studies. It is a rich subject for many disciplines, reflecting as it does some massive social and economic changes in marriage, the family, and procreation, all of them still underway.

One of the major changes is of course the delay of marriage and procreation in developed countries. The average age of marriage has risen (now close to 30), the first child (if any) comes later, and the decision to move from childlessness to one child, and then to two, and then to three (for some) becomes more complex. Later procreation has become the rule with educated women. Would it, however, be better for women to return to earlier procreation (in their twenties rather than thirties), and could that be done with no drastic harm to their professional and work aspirations? I believe it would be. To make that case, I will look at a number of reasons why women have come to delay childbearing, what some of the benefits and harms are, and what it would take to bring a change to earlier procreation.

D. Callahan (✉)
The Hastings Center, Garrison, New York, USA
e-mail: callahand@thehastingscenter.org

F. Simonstein (ed.), *Reprogen-ethics and the Future of Gender*, International Library of Ethics, Law, and the New Medicine 43, DOI 10.1007/978-90-481-2475-6_8,

Some Personal History

Except for a number of demographers and some sociologists, comparatively few men have entered this territory. Since men and some of their attitudes and gender folkways are part of the problem (and for a few as we shall see *the* problem), let me say a little about why it came to interest me. The reasons are both personal and professional. On the personal front, I was born in 1930 and came from a Roman Catholic family of two children. My parents, as well as uncles and aunts, had their children during the Great Depression years of the 1930s when birthrates were very low. My father was one of 11 children, but out of nine marriages in his family there were only seven children. My mother, an intelligent woman and an acclaimed beauty, was a traditional stay-at-home mother whose education ended with high school. She seemed to have no aspirations to do outside work, but she did become a prize-winning painter. In our era she would surely have had a professional life. My father was a journalist, one who preferred to drink and play poker with his pals than to stay home.

My wife-to-be, Sidney, was a very different person, surely transitional for the times, the mid-1950s. She went to Bryn Mawr College, a women's college dedicated to developing intellectuals and professionals. Its founder in the late l9th century, M. Cary Thomas, reputedly said that "only our failures marry." My wife came out of that college with two aspirations, one of them to have a large number of children, and the other to gain a Ph.D. in psychology. She wanted both to marry and to have an academic career. I thought those were fine goals and, against the advice of some male friends – wary of women intellectuals – married her at age 24 (she was 21).

We went on to have six children and only later discerned that we had been classic Baby Boom parents. My wife was born in 1933, a cohort who in that year had an average of 3.8 children, the highest in the 20th century. She went on, while raising our children, to get her doctorate. A teaching career and the publication of ten books followed. Her first books, published in the late 1960s and early 1970s, were on feminism and working women, while all of our children were still at home [1, 2]. In her book on working women she wrote about what she was experiencing, and here I am coming back to her topic, 37 years later. We have been talking about the problem for a long time.

Doing everything she did was a hard struggle, but it worked. With the exception of a few lapses, I think I was a helpful husband as she tried to manage it all. It was surely all the more hectic because five of our children were rambunctious and not-well-disciplined, or always law-abiding, boys. Our life was not helped by the fact that I was in the late 1960s starting The Hastings Center, a research institute. My life was stressful at the same time as hers, not a felicitous combination.

I received a Ph.D. in philosophy but was drawn to issues of ethics in the real world rather than an academic career. In 1969 I was asked to spend a year at The Population Council in New York to examine the ethical problems of family planning programs and, in particular, the vigorous efforts then underway to reduce birthrates in developing countries. I was already known as a one-time Catholic who agitated against the Church's rejection of contraception. The initial efforts in the population field focused on making contraception and abortion available, but by the 1980s there was a major shift –brought on by some important UN conferences on women and procreation – to an emphasis on empowering women as the key issue in population control efforts. I spent another decade working on those problems but then moved along to other issues.

My interest in demography, procreation, and family life was rekindled a few years ago by newly emerging evidence that in most developed countries (save for the U.S.) low birthrates, and a proportionate increase in the elderly – technically known as a change in the dependency ratio – posed an economic and social problem. No one thought of that possible issue in the late 1960s. More or less simultaneously our home life was changing as well as our children moved into adulthood. It soon became clear that our children (now ages 42–52), like most of their peers, were slow in getting married (but fast in having "relationships"), and showed no strong interest in having children – even though they all agreed that it was wonderful growing up in a large family and they remain close to each other. Three of them are now married and have a total of 5 children among them – back where my parent's generation was in the 1930s!

In 1996 a daughter-in-law, age 35, died a day after giving birth to a girl. She died from the complications of a c-section, a danger I only learned later in that arises with an increase in maternal age. We brought my son, a screenwriter, and our granddaughter home with us from California, and they have been with us ever since. We fully share responsibility for raising her, and found out was it was like to begin the duties of parenthood all over again in our mid-60s – not easy or recommended, but perfectly possible. Besides the delight we take in our granddaughter, we have had the added bonus of getting to know, through her playmates, a wholly different cohort of parents and through them many insights into the present status of women, work, and childbearing.

Collected Observations

This has probably been an over-long introductory story, but may give some sense why this issue has for me more than passing professional interest: it has joined my professional and private life in stimulating ways. Many observations over the years have led me to think that early procreation is better than late procreation, and that a decision by a woman to delay childbearing until well established in a work or professional role has more burdens than benefits.

Three observations came into the foreground. One of them was to learn about, and to see firsthand, the medical hazards to mothers and babies born after delayed procreation (particularly over age 35). Another is the psychological and often anguished problems of women desperate to get pregnant after a long delay. Still another observation is the considerable difficulty women have in finding a suitable mate after delaying a search until professionally secure. Some other, less noticed, issues arise also – the so-called "sandwich generation" problem, as well as a developing trend for American professional women to drop out of work to stay at home with their children – which I will come to below.

I can well understand that many women feel they have no real choice about delaying procreation. Their own professional maturation and long-term financial security require them to give temporal priority to their career development. Even when they know the many pitfalls and problems for doing so, they feel it is a necessary tradeoff. I know countless women who have done just that and, for most, the outcome has been satisfactory. They pulled it off, but often with considerable travail along the way.

The necessity of that kind of trade-off is not a good situation, much less one that should continue indefinitely. There must be a better way, a different kind of trade-off, if only because many of the consequences of late procreation together with assorted efforts to get pregnant and to have a healthy baby are, by any estimation, harmful. A return to early procreation seems to me matter of maternal common sense – if it can be achieved without drastic harm to professional aspirations.

How did women get in the present situation? There is a large demographic and sociological literature in response that question, far too much to set out here, but there is general agreement on the main lines of an explanation. While the education of women in large numbers began in the mid-19th century, and particularly in higher education, it accelerated rapidly after World War II. Since then the pace of female education has steadily picked up, so much so that there are now in the U.S. and most developed countries more women than men entering, and graduating from, universities and professional schools. While some women have for decades been forced to work, an increase of professional opportunities made a self-chosen work life more attractive. Simultaneously, divorce rates were rising and women came to see the necessity of having an independent income if their marriage did not work out.

Most important, women who wanted both a career and children increasingly came to perceive that procreation had to be delayed to make that economically possible. Social and welfare policies in most countries responded erratically to this situation by enacting laws and regulations to facilitate work and family life. But there has been an enormous variation in such policies, with some very good (France, Sweden) and others poor (the U.S., Spain, and Italy) [3]. These differences have affected not only the working conditions of women but also birthrates: France and Sweden high, Spain and Italy low. In all the developed countries, however, women's age at the birth of the first child has increased, and the more educated a woman is the later that happens.

The Hazards of Delayed Procreation

Over the last four decades or so there has been a significant increase in the proportion of older women bearing children. There has been a doubling of those bearing children after age 30, a tripling of those over 35, and a four-fold increase of those over 40 [4]. With increasing age there is a parallel increase in prolonged and dysfunctional labor, excessive bleeding, breach birth and caesareans (a function of prolonged labor), infertility, a decrease in mean birth weight of infants as well as premature and very low birth weight babies, chronic hypertension, pregnancy-associated hypertension, maternal diabetes, spontaneous abortions, and infant death. Assisted reproduction technology (ART) can contribute to those hazards, though with less certainty about the relationship between age and the use of that technology [5–7].

To all those risks may be added that of multiple births associated with fertility treatments: twins most commonly, but triplets and quadruplets as well. As a friend once observed to me, "I live in an affluent section of the upper west side of New York where the typical parents have grey hair and are pushing twins in strollers." As I was walking with my granddaughter a few years ago, a woman said to me "are you the father or grandfather, you can never tell these days."

One of my daughters-in-law, age 40, gave birth to a son after a long struggle with IVF (including a number of miscarriages), and her experience is one of the statistically few happy stories. The success rate in the U.S. at the best infertility centers is rising, but still low, about 25%, and the costs high, sometimes well over $50,000 for a series of cycles. There is considerable debate about the infertility industry in the U.S. It is wholly for-profit, lacking in government oversight and regulation, and has been charged with fostering unrealistic hopes and engaging in assorted dubious practices [8].

An interesting French study of ART found that it can make up for only half of the births lost by postponing a first attempt at pregnancy from ages 30 to 35, and less than 30% after age 35 [9]. The study concluded with two recommendations: "The message for a woman [over] 35 years trying to conceive is: be patient. . .more than half of those still childless after 1 year will conceive during the next two years. The message for women over 35 years is: "be impatient. The chances of rapid spontaneous conception are still significant, but in case of failure, ART will not fully compensate for the years (and the chances of conceiving) lost." In sum, women's fertility begins to decline in the late 20s and substantially by the mid-thirties. ART can help some women overcome that biological reality but by no means all, and by no means without some significant hazards for mothers and babies when they try. Men's fertility, by the way, declines significantly by the late 30s [10].

The last of our six children was born when my wife was 32 and I was 35; and that pattern was common in the 1950s and 1960s. By then our problem was to stop all that all-too-easy procreation – which seemed to be achieved by the two of us simply breathing the same air – effectively achieved by my wife's sterilization. Shortly thereafter she began her Ph.D. studies, but had written two books before that with three of our children still in diapers. When asked how she managed all this, her usual answer is "I don't know, it must have been someone else."

The Sandwich Generation and Male Fecklessness

Two other considerations are worth introducing at this point, on the "sandwich generation" and on male fecklessness. My wife is now the legal guardian of her mildly demented stepmother, now 95. She thus missed being part of the growing phenomenon of the "sandwich generation," that of women responsible for the care of an elderly parent while still responsible for the care of their young children. My wife was lucky, old enough and with time enough to pay serious attention to her stepmother.

But imagine a scenario that will become increasingly frequent, that of a woman who has one or two children after 35 and a mother age 70. The mother at that age will probably not need assistance, but she almost certainly will when she is 85 or 90. At that point, her daughter's children will be between 15 and 20 years old, still under her care. That situation will create a far more severe sandwich generation problem than my wife had to face. With the fastest growing age group being those over 85, and the large number of women procreating after 35 that is a perfectly predictable scenario. While much earlier procreation might not obviate the sandwich generation problem, it would surely ameliorate it.

It would be naïve, moreover, to think that men in the future will be any more eager than they now are to be caretakers of elderly parents or parents-in-law. Some surely will, but most will turn it over to their wives. There have been long-standing complaints about this imbalance of responsibilities. Yet if one considers the fact that female life expectancy is on average about seven years longer than men's, it is most likely that the elderly female family member is the one who will need help. In my experience, elderly women would prefer to have women care for them rather than men, particularly if they are frail and need intimate assistance. Women are in effect drafted to care for women, but by other women. This does not make it any more fair for female caretakers, but does help explain the persistence of the pattern.

The last problem of late procreation I will mention is that of male fecklessness, an age-old pathology. Many of the professional women I know who have delayed marriage and procreation have experienced a now-familiar situation. They delayed marriage but not cohabitation, thus getting some of the benefits of marriage (sex and intimacy) but without the burden of celibacy and babies. Often enough they hoped their partner would eventually want to get married, so they persevered, ever hopeful.

Often enough, however, the partner jumped ship, sometimes to marry someone else, sometimes to find a new partner, sometimes just vanishing [11].

The woman was by then in her mid- or late-30s, knocking at our door to ask us if we knew any eligible men. As it turns out, most of the eligible men in that age group are either already married or the sort likely to habitually move from one short-term relationship to another. Often it took years to find someone who was eligible, and they often failed altogether – and sometimes they only succeeded when they were in their early to mid-40s, too late to have favorable biological odds to have a child.

While it is worth an article in itself, my own observation is that effective contraception and abortion have far more empowered men to avoid marriage and procreation than women. More often than my pro-choice colleagues like to acknowledge, many men coerce women to have an abortion. The feckless male has been a problem for a few thousand years, but it is much easier for them now to evade binding commitments than it ever was in the past. Almost any woman who has put off permanent bonding and procreation until later in life will know what I am talking about.

I have tried so far to lay out the general problems that a woman may encounter by delaying procreation. I use the word "may" because many lucky women are spared them and many others weather them in fine shape. One could simply then argue that life is full of trade-offs and that the advantages of late procreation for women well outweigh the disadvantages. My contention, however, is that it would be better to reduce the probabilities of a troubled or bad outcome by moving procreation back into the 20–30 year range. A different trade-off is possible.

Alternative Posibilities

Before developing how that might be accomplished, it is worth considering an alternative possibility. Why not attempt to radically improve the situation of late procreation, biologically, economically, and socially, getting rid of or at least sharply reducing the pitfalls? It is possible, for instance, that every developed country could have family and maternity policies as good as those in France. That country historically has led the way in developing policies and programs supportive of procreation in general and working mothers in particular. The result is that they have a birth rate which is now the highest in Europe, close to 2.1 children per woman (the population replacement level), lavish paid maternity leaves (for husbands as well), and fine daycare facilities – and all of this in a country with the highest proportion of working women of any European country. Other northern European countries have good policies as well, but the southern European countries – Spain, Italy, and Greece – have poor ones. The latter countries, moreover, have some of the lowest birthrates in the world, few possibilities for women to work part-time and poor maternal leave and childcare policies. The United States has a birthrate at the replacement level, heavily due to immigrants, but poor maternity leave and day care policies.

In principle, the countries with poor policies could change, but that would be of little help in dealing with the medical problems of late procreation, even common in those countries with good policies. In any case, the countries with weak policies have been slow to change and, though fully aware they have problems, many of their cultural values – especially traditions of family rather than government support for marriage and maternity – work against a serious shift.

A Failure of Feminism?

In 2005 a prominent American feminist, Linda R. Hirshman, wrote an article that stimulated considerable debate. "Half the wealthiest, most-privileged, best educated females in the country." she wrote, "stay at home with their babies rather than work in the market economy," and noted that their proportion was increasing [12]. "Feminism among the educated elite has "largely failed in its goals. . ..there are few women in the corridors of power, and marriage is essentially unchanged. . ..private lives have hardly budged. The real glass ceiling is at home. . .women are letting their careers slide to tend the home fires." She scoffed at the reasoning of some feminists, who argue that women should be free to choose whether to work or stay at home. Those who chose the latter, she more than implied, have simply been co-opted into accepting traditional male-female divisions of labor.

For Hirshman, the tragic part of all this is that a life for women focused on the family "allows fewer opportunities for full human flourishing than the public sphere like the government or the market." The real glass ceiling for women, she argued, is not out in the workplace but at home, caught in the conservative trap of domesticity – but foolishly thinking they had made a free choice. Her solution is not better government policies for working women and working mothers, but a change in the women themselves: aim for careers where money is best made, for money is power; move away from an education in the humanities toward courses in economics; and have, at most, one child.

I found her article strange and unperceptive. One of the worst features of American society among the educated elite is a worship of money, a crass materialism. As a colleague of mine, fixated on a high income as a goal of life once said, when I told him that one of my sons wanted to be a high school teacher, "he's a smart boy, why would he want to do that?" Her own article reveals some of the real motives of many of the very drop-out women she interviewed, and then castigated. They did so, she reports, not because they were trapped in traditional male-female roles (none she cites mentions that), but because, after sampling that great and glamorous outside world of money and power, they quit because they "were already alienated from their work or at least not committed to a life of work. . . .even where employers offered them part-time work, they were not interested in taking it."

Hirshman seemed not to notice that her own examples refuted her argument. Her informants did not blame men at home, but the nature of the professional work, too often boring and stressful. As the late black writer James Baldwin once famously noted of white society, "who wants to be integrated into a burning building?" Many educated women, having tasted the fruit of work in Hirschman's more "fulfilling" society, choose not to be integrated into it. Hirshman's opening line about women who "stay at home with their babies," is faulty in two ways. One of them is obvious: babies do not remain babies for long. They grow up and it is fulfilling to help them do so. The other is that every drop-out mother I have known does community work, serving on a school board, running for political office, doing part-time work of one kind or another, painting or writing novels. The changing of diapers comes and goes quickly. It is their home and their local community that draw their attention and energy. They think of themselves just as self-fulfilled as those who concentrate their efforts in the marketplace.

But whether their focus is on life in the local community or the larger one, their delay of procreation – which came about because they gave their initial priority to getting their careers underway – carries with it the same hazards and liabilities as those who stay in the public workplace. Indeed, there is sometimes the added irony that, if they decide to stop work to stay home in order to have a third child, they may well have trouble doing so because of their advanced age.

In any case, Hirshman's denigration of family life and her beatification of life in the market and government, offer no serious solutions for women who want to combine family life and a career, nor is it of any help in dealing with the cluster of delayed procreation problems. Harriet Bradley's article (Chapter 9) nicely brings out the harm done by "contemporary economic rationality, the logic that springs from capitalism." Linda Hirshman seems to offer a feminist version of that economic rationality, but I think it is a dead end, good neither for women nor the rest of us. For that matter, the French solution of improved social and welfare policies for working women is of no help to their procreation problems if their desire to work leads them to delay it. Every country should have policies like those of the French: they are fine and necessary, and reflect a seriousness about family life utterly lacking in Hirshman's analysis. But that solution deals with only one aspect of the problem.

An End to Misplaced Tolerance

Work life and public policies should be organized to make earlier procreation possible, and with a trade-off not worse than the one that goes with late procreation. Were it not for the fact that women now correctly perceive delayed procreation, whatever its problems, as the price of professional success, no society would otherwise tolerate them. In any other context low birth weights, undesired multiple babies, and assorted other threats to the health of babies (including death), and for the mother hypertension, diabetes, multiple miscarriages, and infertility would be intolerable. In this case, however, there are no serious efforts underway to change that situation; it is in effect given a pass. The general attitude seems to be, "well, that's just the way it is."

Is change possible? Any change must above all be based on a solid foundation of public policies that support marriage and the family, and no less (1) equally support the aspirations of those women who want work and a career only, (2) those who want only to stay at home, and (3) those who want to mix motherhood and work. In addition to that foundation, some specific changes will be necessary.

First, women will have to understand the value of early procreation, and be willing to resist social pressures that push them toward later procreation. While I have seen no surveys on the subject, my impression is that, while women understand that late efforts at procreation decrease the odds of success, they are far less aware of the other hazards they may face. Second, the workplace will have to accommodate women who procreate early and prior to a professional career on the same basis of equality that it now accommodates those who start their professional life early and procreate late. Third, women will have to look for permanent mates while still young and gaining their education, and seek out those males who will be capable of helping them as two equals with a common goal.

The Importance of Early Procreation

I have presented above information on the medical and social hazards of delayed procreation. They can be called probabilistic harms, that is, the later procreation is delayed the greater the likelihood of medical harms to mother and child. But harms based on population probabilities are by no means certain harms to individuals. Smoking increases the likelihood of lung cancer and heart disease, but many lifelong smokers do not contract either of them. So too with late procreation: it increases harm but does not guarantee it. A major problem for public health policies where risk of harm is present, but not certain harms, is to persuade people to pursue good health behavior when a failure to do so may not harm them. Hence, there are many women who understand the risks of late procreation but choose it anyway out of a sense of necessity, hoping that they will be among the lucky ones who will sail through it all; and many do. Anna Smajdor (Chapter 8) usefully reminds us about the difficulty of determining the biologically optimal time for motherhood, which may not coincide the socially or individually optimal time in life to bear a child.

Even so, I imagine most women would prefer not to run the risks at all if they could achieve their career aspirations without doing so. It may thus be easy to persuade them that they would, all things considered, be healthier mothers with healthier babies if they procreate earlier. They may well agree about that, but it will be insufficient. They must also be persuaded that they will not suffer long-term professional harm for doing so. The possibility of a health risk to mother and child alone is not likely to be persuasive. Only significant changes in the job market will make a difference to women worrying how to mix motherhood and work.

Changing the Job Market

Most jobs in most businesses, organizations, and universities have a built-in rhythm. They take in young workers to work at the bottom of the professional ladder and expect them over time to work their way up it as times goes by. If their work has gone well, by their forties or fifties they will have advanced to leadership positions. It is not necessary these days that the workers move up the ladder at the same organization where they started, but they are expected to have the necessary experience on a ladder such that they can make lateral moves to other organizations. Most organizations, moreover, have been heavily oriented in the apportioning of responsibility and preparation for leadership positions to full-time positions. Few adjunct and part-time professors, male or female, are ever directly elevated to be full tenured professors, just as few part-time corporate worker make the jump to CEO.

I do not believe that any of those organizational structures were deliberately designed to exclude women or to keep them out of leadership roles (though bias against women, simply because they were women, often appears in the climb up the ladder). Actually they make considerable sense, allowing the young their chance to replace the old but demanding in turn that they serve their apprenticeship. It is a structure perfectly adapted to the biology of males who, to procreate, need no time out from work, no adaptations needed for the biological duties of fatherhood. It was designed by men for men, but with no explicit aim of keeping women out. Obviously as well, it no less a structure ill arranged for women, whose biology imposes a different rhythm on their lives if they desire as well to have children. Understood this way – not as a conspiracy against women, but as a structure designed for a modern industrial society – it will be hard to change in any significant way. It requires a long and often demanding apprenticeship, rewards those who most single-mindedly work their way up the ladder, and gives the top prizes to those who beat out everyone else in the intense competition to come out ahead.

Policies can and have been developed to improve the working lot of mothers, as France and the Scandinavians countries show, but that is not the same as putting women on an equal footing with men. Delayed procreation can close the gap and it surely helps, but often at a high medical and psychological price. It can be furthered closed by policies allowing part-time work and work from home – but there will always be tension in trying to serve two masters at the same time. The demands of home and the workplace are different and probably always will be. Conflict is inevitable. Many women can beat the system but by no means all of them. Ironically, I note, an American study of two-income families found that they are financially more vulnerable than single-income families. The basis of this counter-intuitive finding is that such couples often have a level of expenditures (particularly high mortgages) that requires two incomes – and they are thus more vulnerable when one of them gets sick or loses a job [13].

There are two ways to go in balancing work and procreation. One of them is to further nourish the present efforts to find a via media: continued late procreation, improved ART for women who have trouble getting pregnant, increased possibilities of taking time off with as small a loss of upward mobility as possible, strong government support policies, an acceptance of inevitable strain in trying to serve two masters, and a curbing of ambitions to become a CEO or to fully compete with males for the top academic prizes and positions. If all goes well at work and home, there will be sufficient fulfillment to satisfy most women. They will know what stereotypes of motherhood usually obscure: the period of crying babies and sleepless nights is only a short time, but the joy of watching a child grow into maturity is a long-lasting pleasure. In short, continue to muddle through, hoping to improve the combination of work and motherhood.

An Alternative Track

I think there is another possibility, at least in theory. Keeping in mind the need to avoid the present hazards of delayed procreation, two tracks could be imagined. One of them I just sketched, muddling through in an improved way. The other would be a second track, allowing women to gain their professional training, put in a short apprenticeship time, and then work part-time for a decade or so to have and raise their children before returning to full-time work. This track would lead to a strong and striking decline in the problems of delayed procreation, which could mainly be complete by the early 30s. No less strikingly it would help to deal with a much-discussed obstacle to women's professional mobility, the fact that many drop out of the work force, making them less reliable workers than males, who rarely do that.

Those in this new track would have their most difficult child raising years behind them, and they would thus be more likely to last, an attractive asset for an employer. The employers in turn would have to develop part-time positions, designed to achieve a variety of goals: keeping women in the program informed about activities and development in the organization, giving them substantive part-time work of a kind that could be done at home, and providing special day care and other assistance to allow the women to periodically spend short periods of time with the organization to keep personal ties and familiarity strong.

Much would depend upon the willingness of employers to accept and actively support such a track. It would work best with a well publicized and formal designation, known to employers and carrying with it an explicit promise that dropping out (for maternal reasons) after the interim period would be rare. After working for a time, women could apply for the long-term maternity track, with the understanding there would be no prejudice against them for doing so.

It is reasonable to surmise, however, that women who choose that track will not do as well in going up the professional ladder as those who work full time and choose to accept the problems of late procreation – though the latter may be more likely to drop out altogether. The issue thus comes down to the best kind of trade-off to accept. Late procreation has medical and other problems. Early procreation will have more professional problems. But women should have the option to make that choice. Some women de facto already do that, hoping for the best. A formal and recognized dual track system would improve their range and legitimacy of choices.

Men: The Eternal Problem

While both of the tracks I have sketched allow women to find their permanent mates early if they choose, the first track tends to discourage them from doing so: monogamy is often delayed along with childbearing. The alternative track I am suggesting would provide an incentive to seek a mate early, when the pool of unmarried men is much larger than what will be available after the age of 30. It is not only the larger number of men is available earlier, but a larger number proportionately likely to be supportive of a working mother.

I suggested above that many men take advantage of women who want to delay marriage and procreation. They often run from a relationship with women who decide the time has come to pay attention to the clicking of the biological clock. That's their alarm bell to leave. Finding a male who will not do so becomes then all the harder; they have already been taken.

There is of course an obvious problem here. How can one know that finding a compatible and marriageable male in, say, his mid- late-20s will work out better than finding one in his mid-30s? There can be no certain answer to that question, although there is evidence that men and women who cohabit prior to marriage have a higher divorce rate than those who do not. All things considered, a young woman would be wise to look for men who do not have a history of transient relationships, more likely with a male who is simply not old enough to have such a history. For too many men, short-term relationships become a habit, hard to break. Women should stay away from them altogether; they are likely to be the losers.

I wish, finally, I could say something hopeful about the direction of the contemporary male. More of them than in the past are willing to be good helpmates to their wives, and some few will be willing to put their own careers temporarily on hold, or to fully split their professional time to give their wives a chance. But for them, as for the women who might choose my alternative track idea, it is not easy for a man to do that without jeopardizing his career to some extent. More men are working from home, thanks to internets and other technical advances, and some are in industries that are open to part-time workers. In the end, however, modern work life is not congenial for working mothers or working fathers, and that is all the more true if one chooses fast-moving, demanding lines of work. Those jobs pay the most, have the most prestige, but are usually stressful and hard on family life. Men no less than women should, if they are serious about a family life, be careful to choose those occupations and professions that offer flexibility and seem to take seriously the family demands of their employees. There is a trend in that direction, but there is a long, long way to go.

In the meantime, women will all too ruefully understand that men are still men, for better or worse. I won't say we men can't help ourselves, but there is considerable biology and ingrained acculturation to overcome, and I expect there always will be. As a veteran of the androgyny bubble (circa 1960/1970) – the utopian belief that gender is nothing more than a repressive social construct, soon to be transcended – there is nothing like raising actual boys and girls to get over that fantasy. Trying to improve gender roles and responsibilities while improving the relationship between work and family is long-term task, hardly out of its infancy.

References

1. Callahan, S. (1965) *The Illusionof Eve: Modern Women's Quest for Identity.* New York: Sheed & Ward.
2. Callahan, S. (1971) *The Working Mother.* New York: Macmillan.
3. Myers, M., Gornick, J., Ross, K. (1995) Public Childcare, Parental Leave, and Employment. In Sainsbury, D. (ed.) *Gender and Welfare Regimes.* London: Oxford University Press.
4. Luke, B., Brown, M. (2007) Elevated Risks of Pregnancy Complications and Adverse Outcomes. *Human Reproduction* 22(5): 1264–1272.
5. Dolitzki, M., Soriano, D, Schiff, E., Chetrit, A., Masiach, S. (1998) Effects of Very Advanced Maternal Age on Pregnancy Outcome and Rate of Cesarean Delivery. *Obstetrics and Gynecology* 92(5): 935–939.
6. Seoud, M., Nasser, A., Usta, I., Melhem, Z., Kazma, A., Khal, A. (2006) Impact of Advanced Maternal Age on Pregnancy Outcome. *American Journal of Perinatology* 19(1): 1–8.
7. Ziadeh, S., Yahaya, Y. (2001) Pregnancy Outcome at Age 40 or Older. *Archives of Gynecology and Obstetrics* 205: 30–32.
8. Spar, D. (2006) *The Baby Business: How Money, Science,and Politics Drive the Commerce of Conception.* Boston: Harvard Business School Press.
9. Leridon, L. (2004) Can Assisted Reproduction Technology Compensate for the Natural Decline in Fertility With Age? *Human Reproduction* 19(7): 223–240.

10. Dunson, D., Colombo, B., Baird, D. (2002) Changes with Age and Duratin in the Menstrual Cycle. *Human Reproduction* 17(5): 1399–1403.
11. Eggebeen, D. (2002) The Changing Course of Fatherhood: Men's Experience with Fatherhood in Demographic Perspective. *Journal of Family Issues* 25(4): 486–506.
12. Hirshman, L. (2005) Homeward Bound, *The American Prospect* December: 21–26.
13. Warren, A., Tyagi, A. (2003) *The Two Income Family: Why Mothers and Fathers Are Going Broke.* New York: Basic Books.

Chapter 8
Between Fecklessness and Selfishness: Is There a Biologically Optimal Time for Motherhood?

Anna Smajdor

Introduction

In 2005 a group of fertility experts published an editorial in the British Medical Journal warning of an 'epidemic' of older (post 35) mothers in the UK and other Western countries. The authors urged women to have their children at the biologically optimal age (between 20 and 35) and stressed the severe medical, social and economic costs of later reproduction [1]. Public awareness of the trend towards older motherhood has been increasingly reflected in the media. In 2006, an article in The Times declared: 'Late motherhood as big a problem as teenage mums' [2]. In his exploration of the phenomenon, Daniel Callahan suggests that in fact throughout the developed world, there is a growing tendency for women to reproduce later in life, and he warns of the 'significant health and other threats' that this poses to mothers and children.

Is there really an epidemic of older mothers, and if so, what is the specific nature of these associated threats? RL Shaw and DC Giles argue that 'Prior to the Second World War there was nothing unusual about childbirth at ages beyond 40, and in some societies ... women continued to have children well into their forties and beyond' [3]. What has changed is that women are not using all of their reproductive years for reproduction. They are having smaller families, where they reproduce at all, and they are having these families in the latter part of their fertile years. Yet it is not just older mothers that come in for criticism. The countries experiencing a surge in late motherhood are struggling with a problem at the other end of the reproductive spectrum. 'Underage' pregnancies are viewed with alarm, and the UK's record in this respect has been compared unfavourably with other European countries [4].

Yet while the socially-accepted age parameters for reproduction are narrowing, some research suggests that the window of *biological* fertility is widening. Girls are becoming fertile at a younger age, while the average age at menopause is increasing [5, 6]. This could be seen as a good thing: at the same time that women's social and career options have broadened, their fertility span has widened, giving them greater

A. Smajdor (✉)
School of Medicine, Health Policy and Practice, University of East Anglia, Norwich, England, UK
e-mail: acsmajdor@googlemail.com

F. Simonstein (ed.), *Reprogen-ethics and the Future of Gender*, International Library of Ethics, Law, and the New Medicine 43, DOI 10.1007/978-90-481-2475-6_9,

reproductive choice and control. Or so one might have thought. However women are increasingly being discouraged from using their span of fertility to the full. We may be able to countenance women attending university, and working as doctors or lawyers or politicians. But we still cannot stomach the idea that they may have children either towards the beginning or the end of their fertility span, rather than right in the middle.

Why is it that younger and older mothers are the focus of such concern? Largely because in affluent Western societies, reproduction is now regarded as being a decision. Women can choose when to have children, and they do not seem to be exercising this choice appropriately. There are several questions to be asked here. What is the nature of the choices that we believe women are making when they have a child at 40 or 15 instead of 25 or 30? And should women really be held to account for the adverse consequences of their reproductive decisions? Does reproduction in the case of younger, and/or older mothers fall short of some kind of biological ideal, and if so, how is this ideal identified? Who, if anyone, is harmed by women's poor reproductive choices?

There is a substantial body of literature supporting the idea that motherhood over 35 is associated with increased complications [7, 8]. The problems associated with teenage reproduction are also well documented [9]. Despite this, some would argue for a presumption in favour of reproductive autonomy even if it results in outcomes that are perceived as less than ideal [10]. My defense of older mothers is not based on any such presumption. Indeed, I agree with those philosophers who argue that reproduction is seldom ethically justifiable whatever the age of the parent [11]. Nevertheless, I will argue that women whose reproductive lives fail to fulfill an idealized notion of biological normality are subject to excessive condemnation.

Before exploring these questions, I should clarify that my argument is not intended to apply to women who wish to access fertility treatments. The issues related to provision of IVF, whether for 'older' or even postmenopausal women, are not the same as those associated with a trend towards later motherhood in general. For this reason, while acknowledging that older women are more likely to have problems conceiving, I focus on older mothers who have not necessarily sought fertility treatment, but who have – by accident or design – become pregnant in the latter years of their fertility.

The Biological Optimum and Unnecesary Risk

The arguments employed by Callahan and many others in the context of older motherhood purport to be broadly consequence-based, and founded in objective biological facts. Women are encouraged to reproduce at the pinnacle of fertility in order to avoid damaging their own health, as well as that of children who are born. Choosing to reproduce when fertility may be on the wane, though biologically possible, is, in Callahan's view, biologically suboptimal. And – by implication – women should be aspiring to the biologically optimal in their reproductive lives.

What does 'biologically optimal' mean in the context of women's reproductive choices? There are two possible interpretations. One might assign an intrinsic value to whatever is naturally ordained for our species. Natural law theorists believe that this kind of intrinsic value can help to provide answers to some of the moral questions related to new trends and technologies [12]. Proponents of this kind of approach might argue that if women are designed to have children at, say, 25, they should not break this natural rule by seeking to have children when they are no longer at their biological best. This approach is not explicitly based on its outcomes, but it is often assumed that following these natural guidelines will incidentally yield better results. The difficulties in sustaining this kind of argument in the context of older motherhood are obvious. How can we *know* that women are designed to reproduce between 25 and 30 when they may be fertile between the ages of 13 and 51?

Another interpretation of the biologically optimal is more squarely rooted in the utilitarian conviction that certain practices are preferable *because* they yield biologically better outcomes. These 'better' outcomes associated with biologically optimal reproduction are likely to be connected with lower risks to health for mothers and/or babies.

Callahan's argument appears to be based on the latter form of argument: women should reproduce earlier not because they are otherwise transgressing the rules of nature, but because they are exposing themselves and their children to unnecessary risk. For the moment, I will focus on health outcomes for mothers themselves. Let us agree that among the reproductive avenues open to a woman, the biologically optimal choice is that which avoids unnecessary risk to her life and health. Adopting this deceptively simple approach, all we need to do is assess the risks related to each variable of reproductive choice – such as maternal age, for example – and this will yield an objective answer to what is biologically optimal.

However, Callahan and many others who follow this line of reasoning do not address the fact that pregnancy and childbirth themselves carry significant physiological risks for all women whatever their ages. These risks range from discomfort to problems that can be disabling or fatal. Even in developed, affluent countries, women still die in childbirth [13], and although the mortality rate has decreased immensely over the past century, morbidity associated with pregnancy and childbirth is still high – and often overlooked [7]. It is safer to use contraception than to have a child. If one does conceive, it is safer to have an abortion (within the first trimester) than to continue with the pregnancy [14].

If the biologically-optimal choice is the one that entails the fewest unnecessary risks, women should avoid pregnancy altogether. If they do become pregnant, they should abort at the earliest opportunity. The fact that women are not routinely advised to do this indicates that there is another underlying assumption here. That is, that while pregnancy is intrinsically risky, it is a *necessary* risk.

Implicit Pro-natalism

For any woman considered in isolation, it would be safer to remain childless. But society does not encourage women to remain childless. Women are exhorted to have babies – at the right time. These exhortations are clearly not designed to secure the maximum biological benefit to women, since they would be better off not reproducing at all. Callahan, along with most of the commentators who express concerns about later motherhood, takes it to be evident that women will – or should – have at least one child. Underlying this assumption is an implicit pro-natalism. The fact that pregnancy is taken to be a necessary part of a woman's life means that women can be criticized for 'delaying' the inevitable.

An example of one of the typical phrases relating to older motherhood: '[m]any people are choosing to put off having children until relatively late in life' [15] seems innocuous enough. But this kind of language, including the repeated use of terms such as 'put off', 'delay' and 'postpone' reinforces the idea that women's lives necessarily include childbearing, whether sooner or later [3]. Callahan's discussion of the topic relies heavily on such terms, implying either that he endorses the idea that motherhood is inevitable for women, or that he has not fully considered their loaded meaning. It is easy to pick up this kind of terminology almost unthinkingly; indeed, it can be difficult to avoid it. However, these words and phrases are not neutral or benign in such a context.

Interestingly, but perhaps unsurprisingly, fertility doctors in particular tend to use rhetorical language that emphasizes the inevitability of reproduction, and the inexorability of the reproductive drive [16]. In her discussion of this phenomenon, Ruth Chadwick quotes Patrick Steptoe, one of the developers of IVF: '[i]t is a fact that there is a biological drive to reproduce. Women who deny this drive, or in whom it is frustrated, show disturbances in other ways.'[17] Continuing this rhetorical theme, in 2005, Robert Winston argued that the biological urge to reproduce is so strong that the misery suffered by those whose drive is thwarted is 'worse than cancer' [18].

According to these commentators, reproduction *is* both essential and inevitable. But do women who reproduce late in life perceive themselves as being inexorably destined to have a child? To answer this, it is vital to note that the trend towards later motherhood is not an isolated phenomenon. Greater changes are afoot. With the possibility of rewarding careers, relatively safe and effective contraception, and in view of the still formidable challenges of pregnancy, childbirth and motherhood, many women do not have children at all [19]. Those who do have children may only come to this decision once the 'optimal' age has past. The choice being made by women is not necessarily an evaluation of older against younger reproductive age, but an evaluation of motherhood at 40, say, against a childless life [20]. And many women are choosing the latter.

One of the problems here is a tension between abstract rhetorical assertions, and the application of this rhetoric to individuals. It may be true in the abstract to say that women's lives include pregnancy and childbirth, or that societies rely on women to replenish their citizens, but it is not true to say of any individual woman that she must become a mother, or that her duties include the production of new members of society. Laura Purdy has argued that 'even feminists have failed to focus sufficiently on the pronatalism and other cultural factors that can lead women to unwittingly make reproductive decisions that may not be in their own interest.'[21] Pronatalism is a strong word, but there seems to be no other that will account for an assumption that women's lives must necessarily include reproduction.

However, if one does adopt a pronatalist view, the criticism of older mothers seems to hold far more weight. The choice is then not whether to reproduce at all, but whether to do so at the time associated with the best biological outcomes. Given this, it is far more straightforward to regard the riskier choice as a poor one. One of the strongest arguments against older motherhood is the increased risks it imposes on babies when compared to younger motherhood (but not too young). Other concerns may include the idea that those women who reproduce at biologically sub-optimal times are displaying character traits that indicate their unsuitability for motherhood. Such character traits might include irresponsibility, laziness and ignorance, in the case of underage mothers and perhaps arrogance, selfishness or coldness in the older women. The timing of reproduction may also have a negative social impact. In all these cases, because of the complexity of the subject, and the number of variables involved, it is difficult to specify a 'biologically' optimal time for reproduction independently of social factors.

How the Biological Optimum is Socially Constructed

The most obvious way to establish a biologically optimal time for motherhood is to focus on selected reproductive outcomes, such as mothers' and babies' health and life expectancy. By looking at outcomes across the age spectrum, it should be possible to establish the maternal age associated with the fewest risks. It is fairly well established that older mothers are likely to experience greater difficulties during pregnancy and labour, including gestational diabetes, high blood pressure and increased rates of caesarean delivery [22]. Offspring of older women are at greater risk of being underweight and suffering from genetic disorders [1]. Later motherhood might well be deemed biologically sub-optimal on these grounds.

In order to identify the biologically optimal period for motherhood, it is necessary to identify a starting point as well as a cut-off point. The article quoted at the beginning of this chapter stipulates 20–35 as the most 'secure' age for motherhood. This is based on evidence that health outcomes for teenage mothers and their children are worse than for those who reproduce in the middle of the reproductive spectrum. Teenage mothers are more likely to experience vaginal and perineal trauma [23] and their babies are more likely to be of low birth weight [22]. However, while poor reproductive outcomes for the offspring of older women are commonly seen as being an inevitable biological corollary of maternal age, poor outcomes for the offspring of very young mothers are largely determined by social factors. 'Underage' parents are typically of low socio-economic status and educational attainment, and these are strong predictors of negative health outcomes for mothers and their children [9].

If health outcomes are the primary focus of concern here, it may seem irrelevant whether they are socially or biologically determined. Nevertheless, it is interesting to note that although the biologically optimum age for motherhood in terms of women and children's health outcomes might be 20–35, this is true not solely for *biological* reasons, but partly because in our culture, young mothers tend to be disadvantaged.

A Question of Degree?

Whether one considers social or biological causes, the health outcomes associated with pregnancy at younger, or older ages may seem fairly clear. I have argued that any biological optimum derived from these outcomes must be in part socially determined. However, perhaps this is not of the greatest importance. Callahan argues that in any other context, the risks faced by older mothers and their offspring would be 'intolerable'. What matters is that mothers and children should not be exposed to unnecessary harm. If pregnancy is treated as a necessary risk, it is possible to argue that while not all pregnancies are biologically sub-optimal, some are. The sub-optimal ones can be marked out simply by the greater degree of risk that they entail. Risks increase as the maternal age increases, so a child born to a woman of 36 is at less risk than one born to a woman of 40 [24].

Here is data showing outcomes for offspring born to mothers in three age groups:

Table 8.1 Risks to offspring [25]

Maternal age	Delivery pre 37 weeks (%)	Stillbirth (%)	Delta birthweight <5th centile (%)	Admission to SCBU (%)
18–34	6	0.47	5.81	5.2
34–40	6.63	0.61	6.13	5.33
>40	8.17	0.81	7.63	5.92

Delta birthweight: the degree to which the observed birth weight differed from the expected mean for males and females for each week of gestation.
SCBU: special care baby unit.

I have chosen these figures because this particular study compared outcomes between age groups rather than within a specific age group. Figures for categories where the differences between risks for the children of older and younger mothers were most pronounced have been listed, and I excluded categories that showed little or no difference. The figures show that, as we know, the risks for offspring of older mothers are greater than those in the 18–34 age group.

It is not just the danger to offspring that concerns commentators on the trend towards older motherhood. The effects on mothers themselves have been noted with anxiety. Damage to older women during pregnancy and childbirth may be doubly worrying, as apart from the risk to the mother herself, any injury or illness will impact her ability to care for her child as well as imposing costs on the economy and health service.

Callahan's description of death of his daughter in law highlights the fact that things can go tragically wrong for women during childbirth. The effects of a maternal death rebound on the extended family as well as on the child who is left motherless. Yet, as the data in the table above shows, there is no obvious age

Table 8.2 Risks to mothers in pregnancy and childbirth [25]

Maternal age	Pre-eclampsia (%)	Gestational diabetes (%)	Emergency caesarean (%)	Postpartum haemorrhage (%)
18–34	0.78	1	8.65	11.24
34–40	0.76	2.85	11.05	14.25
>40	0.79	4.56	14.24	17.99

threshold beyond which reproduction becomes obviously intolerable. Rather, the risks rise incrementally with age. Does this indicate that women are wrong to have children late in life? No amount of biological or statistical data can provide the answer to this question. Some women may find these risks tolerable, and some may not.

Yet perhaps a woman's decision to accept these risks is not sufficient here. Mothers' choices affect others, not least their offspring. For many, this might seem the strongest reason for advocating earlier motherhood. A woman may have the right to put herself at extra risk, but should she have the right to put her *child* at risk? Children born to older mothers are given no opportunity to accept or reject the risks involved in their conception and birth. However, this is equally true of any child. None of us is given a say as to when, whether or how we are born.

When considering the welfare of a future child, the question of risk is especially complicated because the child cannot avoid the risks connected with its mother's age. It must be exposed to those risks if it is to come into being at all. Because of this, a woman considering reproduction after the 'optimal' age may not perceive the danger to her future child as a compelling reason to forego motherhood altogether. And although the risks involved in later reproduction are greater than those involved in reproduction at the 'optimal' age, a woman who reproduces after having passed the 'optimal' age is still likely to survive and take home a healthy child [22, 26].

It is also necessary to recognize that the decision as to whether to have a child is likely to be influenced by factors other than the health risks involved. Maternal and neonatal health are of extreme importance, but this is not all that reproduction is about. Reproductive outcomes may encompass a very broad spectrum of social, biological and other values. These other outcomes also fluctuate with age, in ways that may counterbalance the extra biological risks associated with older motherhood. More than this, when one begins to examine the links between social and biological outcomes, it becomes evident that they are inextricably linked.

What Kind of Woman 'Delays' Pregnancy?

No woman would choose to suffer pre-eclampsia or gestational diabetes, or to give birth to an ailing, premature, underweight baby. Among reproductive outcomes, those related to health rank very highly. But they are not the only factors to be taken into consideration. Happiness, fulfillment, social acceptance, employment status and financial security may be significantly affected by becoming a mother. The belief that parenthood will be fulfilling and rewarding is precisely what motivates many women to become mothers [27]. Whether parenthood *is* ultimately fulfilling will depend on a variety of social and biological circumstances, including the age and socio-economic status of the potential mother, as well as her state of mind.

The characteristics of older mothers are different from those of younger mothers. A common theme running through discussions with older mothers is 'being ready' [28]. This feeling of readiness is a highly complex phenomenon that is intricately bound up with a woman's social, educational, economic and career achievements [27]. In contrast to 'underage' mothers, statistics show that women reproducing after the age of 35 are likely to be well educated [22]. They are more likely than younger mothers to be in employment [29]. They '. . .seek prenatal care earlier, have a healthier lifestyle, and are psychologically and emotionally prepared to deal with pregnancy and child rearing' [8].

It is important to consider what women gain by later reproduction, and what they might sacrifice if they are exhorted to have children before they are 'ready'. It is also relevant to consider the implications for men. Every child is the product of a father as well as a mother, yet fathers' roles as participants in reproductive decision-making are often overlooked [3]. If women are treated as the sole reproductive decision-makers, and they are urged to have their children earlier, this implies that men somewhere will become fathers, perhaps against their wishes. It is highly questionable a) whether this is an acceptable way to treat men, and b) whether this is likely to have beneficial effects on children, mothers and society at large.

Perhaps this is unfair, however. In urging earlier motherhood, Callahan, for example, surely does not intend that women should have children before they (or their partners) are ready. Rather, he believes that women *ought* to be ready at the optimal age. Yet just how are women supposed to get themselves – and their partners – ready? And should they focus on getting ready for motherhood if this involves compromising other areas of meaning and value in their lives?

Callahan suggests that women might feel ready for motherhood at an earlier age if they were not expected to complete their education and achieve their career aspirations first. Similar arguments have been made by fertility specialists concerned about the physiological effects of later reproduction on women's and children's health [1]. If this is correct, then – as Callahan suggests – adapting social structures and expectations to accommodate earlier motherhood might address the problem. Women's youth could then be spent in having children at the biologically optimal time while postponing, rather than foregoing, the benefits of education and a rewarding career. This might seem to be the obvious solution, but perhaps things are not so simple.

Young men have traditionally been portrayed as being reluctant to assume the shackles of marriage and fatherhood. This was partly because they had access to the kind of education, independence and freedom of choice that many women now have. Before settling down to parenthood and the commitments of family life, men wish to explore and experience the variety and richness of life. They also wish to secure their social and economic status and establish their identity as an individual rather than as a parent. So now do women. For many of us, men and women, the desire to have children and the feeling of readiness to do so is partly dependent on having had a fulfilling and diverse experience of life and an opportunity to pursue and attain social, educational and economic aspirations.

In a world that places a high value on education and career success, having a child can represent a dramatic change in status and identity for women [28]. Viewed positively motherhood may be a new or supplementary identity that adds meaning and value to women's lives. But this has to be carefully negotiated. Parenthood may curtail or negate other values and interests – especially for women [30]. It is interesting to consider this with respect to women who have children very young. Motherhood is an identity that one can aspire to and attain with no further qualifications other than being female and fertile [31]. Hence, perhaps, the prevalence of teenage motherhood among young women with poor educational prospects and low socio-economic status. For women who believe that motherhood alone could not currently fulfill their lives, reproduction may not be the issue of primary importance during the ages at which they are most fertile.

Callahan claims that older mothers' reproductive choices spring from economic necessity. However, I am not convinced that this is the case. Perhaps some women do reluctantly pursue educational and career goals, while mourning the passing of their fertile years. However, for others, psychological, economic and emotional readiness for parenthood cannot necessarily be separated from their age and social circumstances. This is a question of values and priorities as well as mere economic necessity. Hence, as Callahan concedes, even in countries where social structures facilitate early reproduction in ways which ought to be compatible with career and educational success, the most highly educated women are still likely to be the ones who have children later in life.

Here is the crux of this dilemma: should women reject whatever benefits are associated with older motherhood, in order to benefit from the biological advantages of early pregnancy and childbirth? Or should they favour social and economic security at the risk of increased physiological complications for themselves and their children? There are surely arguments to be made on both sides here. Perhaps this question would not be so fraught if there were a more obvious threshold at which one could distinguish between tolerable and intolerable risks in pregnancy. But as I have suggested, this does not seem to be the case. The increased risks of older motherhood might motivate some women to change their reproductive plans, but many – especially if not currently sure that they will ever want children – may regard the risks as being fairly negligible in the face of all the risks that pregnancy and childbirth entail anyway.

Selfishness and Mother/Child Conflict of Interest

One of the reasons for people's unease about later motherhood is the notion that women are putting their social interests above the biological interests of their offspring. Perhaps if motherhood is not the defining identity a woman seeks, she should not have children at all. Is it selfish for women to regard motherhood as simply one among many options that life has to offer, and to fit this in amongst their other aims and aspirations?

It is not clear just how far maternal altruism should go. If women tried to follow all the (often contradictory) advice on achieving the best possible outcomes for their offspring, they would have to subordinate many of their own interests to ensuring that their child had the best start. Is this a reasonable expectation? Would a woman who subordinated her life interests in this way be a better mother, or produce 'better' offspring? Once again, there are two possible avenues of argument here. Firstly, if this kind of approach harms children, mothers could be criticized on these grounds. Alternatively, it might be suggested that mothers ought to be altruistic, and that even if their children are not substantially damaged by their selfish tendencies, older mothers are morally repugnant.

Ann Oakley has observed that there is a tendency to 'pose an artificial conflict of interest between women and their fetuses, to remark on an apparent absence of that effortless altruism which is itself a hallmark of femininity' [32]. This claim seems to be borne out by descriptions of older mothers as cold, selfish and calculating, willing to impose unnecessary risks on their offspring in an endeavour to fit motherhood in with their careers [33]. RL Shaw and DC Giles point out the frequent reiteration of 'selfishness' and 'wanting to have it all' in media discussions of older motherhood [3]. The apparent failure of older mothers to put their children's health interests above their own social interests is an example of just the kind of conflict that Oakley describes.

But, as Oakley suggests, this mother/child conflict is a highly artificial construct. In fact, a child stands to benefit from its mother's social, psychological and economic wellbeing just as much as the mother herself. Conversely, being born to a mother who is struggling economically, or who is psychologically unready for pregnancy can have an adverse impact on a child's health that will last throughout his or her life [34]. Children of parents from lower socio-economic backgrounds are at increased risk of asthma and even cot death [35]. 'It is mothers in the poorest socio-economic circumstances who are most likely to experience the death of a child in the first year of life' [29]. Stillbirth rates are twice as high in families where the father is in the lowest socio-economic category as where he is in the highest [29].

Researchers into the effects of social and economic factors on health argue that 'the diversity of conditions of life can somehow become directly embedded in human biology, such that human vitality can be directly affected by social hierarchies', and that 'high socio-economic status is a powerful buffer against both endogenous and exogenous threats to successful human development' [36]. Although the children of older mothers are statistically more likely to suffer neonatal health complications, they are also statistically more likely to be well-off, meaning that they are better equipped to cope with these physiological problems. So too are the mothers themselves.

This being the case, a woman who has a child at 40 after pursuing her economic, educational and other goals, does not reap these benefits at the expense of her child, who assumes all the risks. Rather, the benefits *and* the risks are shared by the mother and child. Likewise, a mother who reproduces at the 'biologically optimal' age shares the risks and benefits of that choice with her child. There are undoubtedly risks and benefits associated with both, but these are highly interdependent, so that it is almost impossible to disentangle the supposed biological risks from those benefits which are purely social. The array of variables that can have an impact on a child's health is staggering. There is even evidence to suggest that being born to parents who have low IQs is associated with poor health outcomes for offspring [37].

I am not trying to argue here that the children of older mothers necessarily do better than those of younger mothers. The point I am making is that the question is far more complex than this. We live in a society that places a high value on goods that are not necessarily compatible with early parenthood. Our children partake of those goods, and also of the sacrifices that we pay for them. To urge women to forego these goods for the benefit of future children is to regard women, their reproductive functions, and their children's biological outcomes, as being separate from the values and interests shared by the rest of society.

Conflicting Ideologies of Motherhood

Given everything that can adversely affect a child not just in the perinatal period, but throughout its lifespan, does it make sense to focus on the relatively narrow question of late maternal age? It has been suggested that much of the concern over older mothers is actually rooted in the 'interests of society masquerading as the interests of the child' [38]. But perhaps there is something more at stake here than concern for children's wellbeing. It is not just issues relating to older mothers that are causing concern, but understandings of motherhood itself. Tensions between incompatible and conflicting ideologies of motherhood place women in an unenviable situation.

Social and pharmaceutical developments may enable women to exercise some choice over when and whether to reproduce. But choice, calculation, prudent planning and postponement do not sit easily with the powerful image of motherhood as an inexorable biological imperative. Motherhood is no longer necessary; it is merely an option. Older mothers bear the brunt of our failure to accept this. Yet women cannot escape public censure by reverting to the old paradigm either. Those who *fail* to plan their reproductive lives, and who get pregnant by accident are also heavily criticized (whatever their ages) [39].

This is partly because unplanned is assumed to equal unwanted [40]. Undoubtedly, some unplanned conceptions result in abortions. Nevertheless, the idea that all unplanned pregnancies are unwanted is an unreasonable conflation. Having a child is a huge step emotionally, physically and financially and many women 'leave it to chance', getting pregnant accidentally, but keeping the child [41]. Nevertheless, unplanned pregnancies are commonly framed as a public health problem [42], and this leaves women with a dilemma. Coolly fitting motherhood in among other goals is frowned upon, but simply allowing nature to 'take its course' is *also* socially unacceptable.

What conclusion can women draw from these conflicting approaches to planned parenthood and ambivalent responses towards maternal age? How can they assimilate these social expectations with the incredible complexity of biological and social factors that will affect their offsprings' outcomes? With the best of intentions, it is almost impossible to identify, let alone guarantee 'optimal' outcomes, whether one focuses on biological or social factors. Yet if women are perceived to go against what is expected of them, they are subject to extreme vilification. In this context, what is surprising is perhaps not so much that women are having children later, but that they have them at all.

References

1. Bewley, S, M Davies, and P Braude. 2005. Which career first? *British Medical Journal* 331:588–589.
2. Templeton, SK. 2006. *Late motherhood as 'big a problem' as teenage mums*. The Times.
3. Shaw, RL and DC Giles. 2009. Motherhood on ice? A media framing analysis of older mothers in the UK news. *Psychology and Health* 24(2): 221–236.
4. Jewell, David, J Tacchi, and J Donovan. 2000. Teenage pregnancy: whose problem is it? *Family Practice* 17(6):522–528.
5. Johnson, MH and BJ Everitt. 2007. *Essential reproduction*. Malden, MA [etc.]: Blackwell, 142.
6. Rodstrom, K, C Bengtsson, I Milsom et al. 2003. Evidence for a secular trend in menopausal age: a population study of women in Gothenburg. *Menopause* 10(6):538–543.
7. Hebert, PR, G Reed, SS Entman et al. 1999. Serious maternal morbidity after childbirth: Prolonged hospital stays and readmissions. *Obstetrics and Gynecology* 94(6):942–947.
8. Tough, Suzanne C, Christine Newburn-Cook, David W Johnston et al. 2002. Delayed childbearing and its impact on population rate changes in lower birth weight, multiple birth and preterm delivery. *Pediatrics* 109:399–403.
9. Schempf, Ashley H, Amy M Branum, Susan L Lukacs et al. 2007. Maternal age and parity-associated risks of preterm birth. *Paediatric and Perinatal Epidemiology* 21(1):34–43.
10. Strong, C. 2002. A framework for reproductive ethics. In Ethical issues in maternal-fetal medicine, ed. Donna Dickenson, 17–37. Cambridge: Cambridge University Press (pp. 32–33).
11. Häyry, M. 2004. A rational cure for prereproductive stress syndrome. *Journal of Medical Ethics* 30:377–378.
12. Oderberg, D. 2005. Towards a natural law critique of genetic engineering. *Philosophical reflections on medical ethics*, ed. N Athanassoulis, 109–134. Basingstoke: Palgrave MacMillan.
13. Deneux-Tharaux, C, MH Berg, M Bouvier-Colle et al. 2005. Underreporting of Pregnancy-Related Mortality in the United States and Europe. *Obstetrics & Gynecology* 106(4):684–692.
14. Grimes, DA. 1994. The role of hormonal contraceptives: the morbidity and mortality of pregnancy: still risky business. *American Journal of Obstetrics & Gynecology* 170(5S) Supplement:1489–1494.
15. Archivist. 2002. Cost of delayed childbearing. *Archives of disease in childhood* 87:311.
16. Malin, M. 2002. Made in Finland: infertility doctors' representations of children. *Critical Public Health* 12(4), 1: 291–308.
17. Chadwick, R. 1992. *Ethics, reproduction and genetic control*. New York: Routledge, p. xvi.
18. Henderson, M. 2005. *Trauma of infertility is worse than cancer, says Winston*. The Times.
19. Bloom, DE and J Trussell. 1984. What are the determinants of delayed childbearing and permanent childlessness in the United States? *Demography* 21(4):591–611.
20. te Velde, ER and PL Pearson. 2002. The variability of female reproductive ageing. *Human Reproduction Update* 8(2):141–154.
21. Purdy, L. 2006. Women's reproductive autonomy: medicalisation and beyond. *Journal of Medical Ethics* 32:287–291.

22. Hemminki, E and M Gissler. 1996. Births by younger and older mothers in a population with late and regulated childbearing: Finland 1991. *Acta Obstetricia et Gynecologica Scandinavica* 75(1):19–27.
23. Briggs, M, W Hopman, and MA Jamieson. 2007. Comparing pregnancy in adolescents and adults: obstetric outcomes and prevalence of anemia. *Journal of Obstetrics and Gynaecology Canada* 29(7):546–555.
24. Cleary-Goldman, J, Fergal D Malone, J Vidaver et al. 2005. Impact of maternal age on obstetric outcome. *Obstetrics & Gynecology* 105(5):983–990.
25. Jolly, M, N Sebire, J Harris et al. 2000. The risks associated with pregnancy in women aged 35 years or older. *Human Reproduction* 15(11):2433–2437.
26. Dulitzki, M, D Soriano, E Schiff et al. 1998. Effect of very advanced maternal age on pregnancy outcome and rate of cesarean delivery. *Obstetrics & Gynecology* 92:935–939.
27. Dion, K. 1995. Delayed parenthood and women's expectations about the transition to parenthood. *International journal of Behavioral Development* 18:315–318.
28. Shelton, N and S Johnson. 2006. 'I think motherhood for me was a bit like a double-edged sword': the narratives of older mothers. *Journal of Community & Applied Social Psychology* 16:316–330.
29. Graham, H. 1993. *Hardship and health in women's lives*. London: Harvester Wheatsheaf, pp. 109, 188, 189.
30. Lee, C. 1997. Social context, depression, and the transition to motherhood. *British Journal of Health Psychology* 2(Part 2):93–108.
31. Hoggart, L. 2006. Risk: Young women and sexual decision-making. *Forum: qualitative social research* 7(1), Art. 28. Section 4.2.
32. Oakley, A. 1993. *Essays on women, medicine and health*. Edinburgh: Edinburgh University Press, p. 84.
33. Foggo, D and L Rogers. 2006. *Fertility experts urge end to "selfish" late motherhood*. The Sunday Times.
34. O'Connora, Thomas G, Yoav Ben-Shlomob, Jon Heronc et al. 2005. Prenatal Anxiety Predicts Individual Differences in Cortisol in Pre-Adolescent Children. *Biological Psychiatry* 58(3).
35. Roberts, H. 1997. Socioeconomic determinants of health: Children, inequalities, and health. *British Medical Journal* 314:1122.
36. Hertzman, C and M Wiens. 1996. Child development and long-term outcomes: a population health perspective and summary of successful interventions. *Social science and medicine* 43(7):1083–1095.
37. Batty, David G, Geoff Der, Sally Macintyre et al. 2006. Does IQ explain socioeconomic inequalities in health? Evidence from a population based cohort study in the west of Scotland. *British Medical Journal* 332(7541):580–558.
38. Hope, T, G Lockwood, L Lockwood et al. 1999. An ethical debate: should older women be offered in vitro fertilization? In *Bioethics: an anthology*, eds. H Kuhse and P Singer, 116–123. Oxford: Blackwell
39. Martino, SC, RL Collins, PL Ellickson et al. 2006. Exploring the link between substance abuse and abortion: the roles of unconventionality and unplanned pregnancy. *Perspectives on Sex and Reproductive Health* 38(2):66–75.
40. Freeman, SB. 2005. Preventing unplanned pregnancy. The right contraceptive fit can make a difference. *Advance for Nurse Practitioners* 13(8):57–59, 61–62, 67.
41. Anon. 2006. One third of births 'not planned'. *BBC News Online* Available at http://news.bbc.co.uk/1/hi/scotland/edinburgh_and_east/6153736.stm (accessed 31st January 2008).
42. Drife, JO. 1993. Contraceptive problems in the developed world. *British Medical Bulletin* 49(1):17–2626.

Chapter 9
Gendered Futures: Reproduction and Production in Women's Lives

Harriet Bradley

Introduction

In the retrospective exhibition at the Tate Modern Art Gallery in 2007 stands one of Louise Bourgeois' cell installations. Wooden panels enclose a bedroom scene. A blood-red mattress lies on a wire bed-frame. A square red pillow (for the man?) sits next to a small white pillow with the words 'je t'aime' embroidered on it in the same scarlet red. Built into the mattress is a penile shape. A child's toy lies on the bed. One of the wooden panels, through which the visitor peers at the exhibit, has a label: 'fermez la porte SVP'. Across the gallery room, a huge wooden vat encloses another cell, this one entitled 'Liquid Pleasures': these, the commentary informs us, are bodily secretions – sweat, tears, sexual flows. Above the bed (again a wire framework) a series of glass vessels or retorts are fixed on metal stands forming a rather beautiful forest above the bed. We are told that this symbolises the bodily release experienced by the adolescent girl as she grows into sexual maturity. But... across from the bed on the wall hangs a huge man's overcoat, perhaps eight or nine feet tall, watching.... In another part of the exhibition a work called 'Do not Abandon me' consists of a flesh-pink knitted fabric woman; she lies on her back and between her legs emerges a knitted child, fastened to her belly by a long pink cord. A whole series of drawings and sculptures throughout the exhibition is entitled 'Femme Maison' (housewife in French). In these artworks, a woman's naked body disappears into a house which becomes her head.

In these pieces, Louise Bourgeois links sexuality, reproduction and motherhood, domesticity and housewifery, bodies and pleasures, through the imagery of entrapment. These and other works offer a disturbing, ambiguous commentary on female experience and identity. The exhibition's symbol is the giant metal sculpture of a spider, 'Maman', who spins her threads from her stomach, as Bourgeois' own mother wove her tapestry fabrics, as the mother in 'Do not Abandon me' dangles her child on the umbilical cord. The spider is huge, unsettling, and its giant legs, too, form a cage, an entrapment.

H. Bradley (✉)
Faculty of Social Sciences, University of Bristol, Bristol, England, UK
e-mail: Harriet.Bradley@bris@ac.uk

F. Simonstein (ed.), *Reprogen-ethics and the Future of Gender*, International Library of Ethics, Law, and the New Medicine 43, DOI 10.1007/978-90-481-2475-6_10,

Bourgeois was born in 1911, when women's experience was firmly set in a domestic subordination mould. Yet, reproduction still stands at the heart of women's lives. Young Western women in adolescence are perhaps increasingly defined by sexuality. Reproductive concerns and the experience of motherhood (actual, desired or renounced) are central preoccupations for most women throughout their lives. Menstruation, fertility, pregnancy and the menopause are bodily realities which deeply affect women and contribute to their construction as medicalised beings. The social valuation of women is still closely linked to their reproductive capacities: their ability to bear children and their sexual potential. Older women may be ignored and made socially invisible because the loss of fertility is associated with desexualisation. In all these ways, the lifetime experiences of women are sharply distinguished from those of men. While fatherhood is an important signifier of masculinity, the fact of being a father does not frame or constrain masculine experience to anything like the extent that being a mother (actual or potential) frames that of women.

Because women's reproductive capacities have an important impact on their productive lives (their ability to work outside the home, the jobs they choose, the hours they work), their labour market involvement remains differentiated from that of men. As Glucksmann [1] states in her conceptualisation of the 'total social organisation of labour', women enter the productive sphere under different conditions and assumptions from men. Thus women's reproductive role lies at the core of persistent gender difference and the inequalities of wealth, status and power that follow. Reproduction is integral to gendering [2].

Is it inevitable that this will be the case? In Chapter 14, Tuija Takala considers a future beyond the polarities of gender. Post-modern and poststructuralist theorists have attempted to deconstruct both rigid identities and gendered bodies to suggest the potential of freedom and choice offered by a framework of plural identities. But in this chapter, I will argue that the future, foreseeable at least, will remain deeply gendered. I will consider how current processes of gendering are set to persist and critically examine possible scenarios for change and liberation.

There is certainly a paradox to be discerned here. As indicated in this book, the period since the end of the Second World War has witnessed a steady development of reproductive technologies, which apparently maximize choices for women, offer control over reproductive processes, make childbearing safer and allow women to programme their life courses in ways to better balance productive and reproductive desiderata. It would seem that now women, in the West at least, can control when, with whom and how they have children. Even more importantly, perhaps, contraception and fertility treatments allow women the choice of being a mother if they want to be or not being a mother if they don't. Yet in spite of these developments women struggle to find a 'work-life balance' (the 'life' part of this formulation interestingly being used by policy-makers in the UK as equivalent to children and domesticity). Indeed in the countries of the European Union (EU) the issue of the 'reconciliation' of family and reproduction needs has become both a major policy focus and one of the most important discussion points in contemporary feminist politics. Nor do the anxieties around reproduction seem to have diminished for individuals. And, if we turn to the developing world millions of women suffer from reproductive toils: over

500,000 die in childbirth each year; thousands of women are raped yearly, with the potential of unwanted motherhood; while other face rejection by their partners if the marriage proves infertile.

Reproduction, then, though bringing many joys to men and women, also looms like a shadow (or Louise Bourgeois' giant male overcoat) over women's lives. Is this going to change? In the next section I consider the processes of gendering as they operate today and then move on to explore some scenarios for the future, which may or may not offer the chance of allaying the dangers while heightening the pleasures of reproductive lives.

What's Going on? The Gendering of Social Life

It should be evident by now that I am using reproduction in a broad sense. I am not just referring to the acts of conceiving and bringing a child into the world (human reproduction) but to the whole nexus of complex social relations that surrounds such acts. This is the notion of 'social reproduction', as employed in classic feminist sociology: the complex sets of arrangements by which societies reproduce themselves, including sexualities, marriage, domestic care, as well as pregnancy, childbearing and childrearing.

In a recent book, *Gender* [2], I developed an account of gender relations drawing on Miriam Glucksmann's accounts of the total social organisation of labour. Using a materialist approach to feminist theory, Glucksmann argues that human work relations cannot be encompassed within an analysis of production alone. She states that for a complete account of gender divisions we must look at the interrelations of production, reproduction and consumption. Each sphere is gendered, resting upon and perpetuating differences between women and men, and the activities in each sphere impinge on the others. Thus, the traditional construction of men as breadwinners in the sphere of production, pushes women to take up the major responsibilities of childcare and domestic labour, seen (rightly) as at odds with the kind of hyper-commitment to career demanded by contemporary competitive capitalisms. The commitment of women to childcare then serves to ensure that they enter into the sphere of production on different terms from men. For example, they are often construed as 'secondary', or 'temporary workers', as being supported by husbands and partners and therefore not needing equal pay; they are often constrained in their choice of jobs by needing to work close to home, or to their children's' schools. For example, in recent work on minority ethnic women in Britain, carried out for the Equal Opportunities Commission (EOC), we found that women from Pakistani and Bangladeshi families liked to work in the school sector, as teachers or educational assistants, or as nursery workers, because these jobs fitted their childcare timetable [3].

In the sphere of consumption, women's association with the home has historically constructed them as prime shoppers, taking responsibility for provisioning the household and for maintaining the status and respectability of household members by choice of consumption goods, such as clothing and home furnishings. But this 'women's world' of home and shopping also intrudes on the spheres of production and reproduction, through the use of female bodies and sexuality as a prime marketing tool. Sex sells! Many 'women's jobs' depend on glamour and sexuality as an attribute of the worker (from models, retail assistants, receptionists and beauticians to hostesses and sex workers of various kinds). The sexualisation of young women then affirms their associations with maternity and reproduction, again positioning them as secondary workers. Moreover, at a deep symbolic level, all the arrangements described above serve to strengthen the association of women with bodies ('supermodel' Elle McPherson was famously nicknamed 'the body'), and of men with mind and brain. Even working-class men, who might be denied the attributes of intelligence by their middle-class bosses and were associated with bodily strength, were classically described as 'hands' [4]. In the symbolic construction of the social body, middle-class men are the heads, working-class men the hands, and women of every class are wombs and erogenous zones: 'tits and arse' to use the vulgar English colloquialism.

Focusing more specifically on reproduction, I distinguished in *Gender* a number of key processes of gendering: the gendered structure of the domestic division of labour which in virtually every nation assigns the majority of housework and childcare to women (typically 70–80%); the ideology of motherhood and maternal instinct, which holds that women are not fulfilled until they have children and too easily categorizes women as 'good' or 'bad' mothers; and the microprocesses of interaction in the family through which children learn gender-specific behaviour. We may add to these the relentless sexualisation characteristic of Western societies which causes women to rate themselves in terms of bodily attractiveness and ability to attract men, resulting in a range of 'gendered' impacts: the spread of STDs, HIV/AIDS, teenage pregnancies, and high profile casualties such as Britney Spears, Amy Winehouse and Lindsay Lohan, who may resort to maternity (often outside wedlock) as a proof of their femininity and to shore up shattered self-esteem. Oddly, the same is true of young women from very poor and disadvantaged backgrounds for whom motherhood may be chosen as a mark of adult maturity, or a way to compensate for failures at school.

There are other forces of gendering involved, however, and if we turn more specifically to pregnancy and childbirth, processes at the heart of reproduction, the key issue for gender analysis is the medical control and management of pregnancy and childbirth. We are at the core of the paradox here. As technology has become ever more sophisticated, safeguards around pregnancy and birth, along with increased ability to control fertility and 'cure' infertility, apparently offer countless advantages to childbearing women and maximise their choices. Yet in tandem with this, pregnant women are more at the mercy of other people's decisions than ever before. They are also more heavily policed. Their diet, alcohol and tobacco intake are monitored. Researchers recently found that drinking two cups of coffee a day may precipitate miscarriage, another warning to heighten the anxiety and curtail the freedoms of expectant mothers [5]. They are weighed, scanned, stripped and poked at. Medical professionals control their access to fertility treatments. They are trained and coached by midwives and health visitors in how to give birth and how to care for their babies. Even though women are allowed to 'choose' their 'birthplan', they do so under heavy medical guidance. If they are addicted to drugs or alcohol, they are designated bad mothers from the outset and may – literally- have their children snatched away from them at birth both in Britain and America. In Britain each year around 1,300 babies under a month old are taken into care and subsequently given for adoption [6].

Thus, Letherby and Marchbanks [7] argue, despite the ideology of 'women's choices', in reality choices about fertility, contraception, childbearing, abortion etc are framed by men – 'medics and partners'. While these events can be interpreted primarily in terms of medical dominance and regimes of power, using a Foucaultian framework, there are clear gender implications. Although men may be encouraged to participate in babycare and attend antenatal classes, they are not compelled to do anything as a result of childbirth, and attempts to police them (in terms of smoking and drinking) have been relatively half-hearted. Thus, in the name of choice, women's behaviour is monitored, and censured if it is found wanting, while men's freedom to behave as they want (bad or good) is untrammelled. In addition, surveillance and medicalisation, while meant to be beneficial by leading to successful births (obviously the desired object), have the side-effect of inducing anxiety and guilt in women, especially if anything goes wrong. A Bristol research project into early miscarriage found that women experiencing miscarriage, and seeking causes for an event which often has no real explanation, readily blamed themselves [8].

According to Inhorn and Balen, women worldwide bear the major burden of infertility [9]. Studies of childless women show that many feel incomplete, or a failure as a woman [8, 10, 11]. In some countries and cultures women who are 'barren' may be cast aside by their husbands. While men who do not have children may suffer teasing and embarrassment about their sexual potency, they are not stigmatised as without value. Masculinity may be enhanced by paternity, but it is not a necessity for successful masculinity. There are other models. A woman without children, however, remains an object of doubt or pity. Woolletts' study of infertile women revealed the social pressure they felt under, especially from their own mothers, to bear children [11]. It is this that induces many women to go through the expense, stress and discomfort of repeated fertility treatments [12]. If women bear the burden of infertility, they also bear a heavy burden from unwanted fertility: each year 20 million women subject themselves to unsafe abortions [7].

To make these points is not to espouse a technophobic position or a dogmatic espousal of natural childbirth. Research shows that women want the technologies of medicalised childbirth and feel protected and reassured by them [13]. We would not wish to return to the conditions that surround birth in the developing world. Reproduction is a dangerous matter As well as maternal deaths, each year millions of women sustain illnesses and injury as a result of complications, 1.4 infants are stillborn and some two million die in the first week of life [7]. But medicalised childbirth poses its own threats. A recent study shows that women who give birth by caesarean section (currently 23% in the UK) have the risk of suffering a hysterectomy after their next birth substantially increased [14].

The problem is not the technologies themselves but the conditions under which they are made available to women and the expectations which frame them. I have suggested in this section that social expectations and imagery of women, what we might call the 'feminine imaginary', are still strongly shaped by the requirements of social reproduction. Thus, for example, Gayle Letherby's research into infertility and involuntary childlessness [15] shows how dominant discourses on femininity and identity constructed these women as outsiders, with no valid claim to contemporary womanhood; her respondents experienced themselves as being excluded from the 'club of mothers' and thus from full membership of society. As Germaine Greer succinctly puts it:

> It is a truism of feminist history that women have been regarded primarily as body, passive fertile body, as essential to human survival on earth [16].

That history has not yet been over-written

Future Positive? Scenarios of Possibility

But are things changing? Yes, to some extent, they are; more women are resisting this reproductive primalism, or attempting to change its rules. However, it is only some women who are able to do this, and only women in some countries. Degrees of freedom and autonomy from male control are needed to achieve change, as well as appropriate economic resources to allow for choice, and many women in poorer parts of the world do not have access to such freedoms and resources. In the remainder of this paper, however, I will focus on women in the highly developed capitalist world, and consider a number of possible scenarios for the future of gendering. First, I shall explore three possible developments which might remove the constraints that reproductive responsibility place on women, especially in relation to their access to equality in the world of paid employment. But I shall conclude with two more scenarios which realistically seem more imminent, and suggest that these more likely outcomes will see current processes of gendering extending into the future.

Scenario 1. Equal Shares

Both women and men are seen as responsible for looking after and raising children. Women and men have equal parental leave rights (no distinction between maternity and paternity leave). This is not just an economic matter, as ideas about looking after children have also changed and the assumption that doing childcare and domestic work is 'women's labour' has ended. Both girls and boys are brought up acquiring knowledge of caring for babies and are taught to cook and clean. Both men and women look after children in nurseries.

Long ago Shulamith Firestone [17] declared that the only solution to gender inequality was a radical shift in relationships within the family. Notoriously, Firestone saw women as voluntarily enslaved through love, and suggested that equality would prove chimerical until children could be produced external to women's bodies, an idea notably taken up in Marge Piercy's Utopian/dystopian novel *Woman on the edge of time* [18]. However, apart from growing embryos in artificial wombs, Firestone also spoke of the need to ensure that relations within the family should be changed to promote equality and democratic decision-making, altering the balance of power between adults and children, women and men. Subsequently, one of feminism's central goals has been to demolish the domestic division of labour so that the labour of caring for children, household and home is shared between the partners.

In Marge Piercey's Utopia, Mattapoisett, each child has three 'mothers', who may be of either sex. Thus the task of mothering becomes a social not biological assignment. Such an arrangement is based on the conception that the task of reproduction is for the benefit of the whole society and consequently the responsibility and labour should be shared by all society's members. Such collective experiments have been elaborated in Utopian or socialist communities, real as well as fictional. But such an approach to social reproduction, although not without its echoes in the child-rearing customs in tribal societies or some societies of the developing world, is at odds with the individualistic frameworks of the advanced capitalist nations which have fostered privatised and social isolated family structures [19]. However, an ideal of shared parenting can be adopted within the familiar nucleated norm. Indeed, such an ideal is often expressed by young people today. Wetherell et al.'s research among students found that women and men espoused egalitarian ideals, and expected to share housework and childrearing in their future marriages [20].

We know, anecdotally and from observation, that such arrangement prevail in some families. We know also that there are small numbers of 'role reversal' families in which husbands stay home and look after children [21]. However, statistically such cases are not the norm. Given the circumstances of today's world and the realities of contemporary work arrangements, shared parenting remains an illusive goal.

Esther Dermott's studies of the 'new fatherhood' [22, 23], for example, reveal that though fathers are more involved with their children this does not amount to equality. Men tend to 'cherrypick' certain tasks. They prize quality-time with their children (play, bedtime stories and trips to the park or pool), but are still less involved with everyday maintenance tasks, bathing, feeding and nappy-changing. Dermott suggests that fathers only do about 30% of caring for children. Although surveys repeatedly report that fathers would like to spend 'more time' with their children (and children with their dads) such aspirations do not translate into a shift to shared parenting. The comment of one of Dermott's respondents is deeply indicative of how fathers view their role in relation to children: 'the real family time is at the weekend, it is unorganised but always the four of us' [22]. But children are not only there at the weekend! By contrast the ideals of committed motherhood, 'total motherhood' as Badinter [24] called it, involve 'being there for them' on a twenty-four basis. Such differentiated views of mothers' time and father's time are still at the core of reproductive arrangements showing how strongly gendered these remain. Two other quotations from Dermott's 'intimate fathers' encapsulate this gendering process

> I see my main activity in the evening, when I may or may not help with the bath or I may or may not help in various other ways, but most evenings – which is very important to me – at their different bedtimes, having read them a story . . . give them a cuddle and talk to them for five or ten minutes.
>
> I would see them for an hour, hour and a half before they go to bed, when they are given the chance to jump on me, and play with my rugby ball and kick that around [22].

As Dermott concludes, the fathers' view of what their roles are and should be construct a potential differentiation between 'intimate fatherhood' and 'intensive motherhood' which is becoming the modern ideal. Couples may be perfectly contented with this. But it does not equate to a degendering of parenting. In chapter 7 Callahan deals with what he calls the 'fecklessness' of many men, which makes them resist taking their full share of such burdens. As Smadjor points out (chapter 8), young men like to enjoy the variety and pleasures of life before they feel 'ready' to settle down. Moreover, as we will explore subsequently, the conditions of contemporary work make it difficult for both partners to work full-time, unless they are rich enough to afford full-time child-care help.

Scenario 2. The Technical Fix

Fertility treatment and control have continued apace, so that all couples can choose if and when to have children. Surrogacy has become increasingly common, allowing older women, and those who are for various reasons unable to undergo the physical experience of childbirth, to have a family. Teenagers are supplied with contraceptives and taught to use them as they reach puberty, to prevent unplanned early pregnancies. The government slogan is 'every child a planned one' – and a perfect one. Medical advances mean that children are born without handicaps or birth defects, thus removing some of the hardest and most time-consuming parenting tasks. Fully planned parenthood means that couples can have children at the most convenient points in their careers. Teleworking and home-based offices with teleconferencing facilities are now commonplace, allowing parents to work mainly from home during the child's early years. Ever more efficient robots make domestic work less of a burden.

Technology to the rescue! Firestone's vision of extra-uterine childbirth, which is explored fully and endorsed in Tuija Takala's chapter, seemed mere science fantasy at the time, but advances in medical and engineering sciences, such as cloning techniques and nano-technology, make the chances of such techniques being developed less unlikely. Takala believes ectogenesis could promote equality. Even without laboratory-grown babies, feasible developments would probably aid couples into moving towards the *Equal Shares* scenario discussed above. Cyberpunks and cyberfeminists present technology as moving towards release from the constraints of the body.

However, it is important to stress that technology cannot *in itself* determine the shape of things to come. It all depends on how it is used. There is a fierce debate within feminism between those who see NRTs being used by men to control women and reaffirm the primacy of maternity and those who believe the new techniques maximise choice and open up freedoms to women.

At the moment I find myself more on the side of the pessimists. If every woman **can** bear a child, will not the tendency to label those who resist as 'abnormal' and 'incomplete' become more marked? However, whichever of these positions is espoused, the reality is that at present these techniques are too expensive to be available to all. Ectogenesis might be a solution in the future, but at present remains a vision and one that might be highly contested by those who believe in the primacy of the natural. In terms of more immediate technical aids, those higher up the occupational pecking order, and therefore subject to relations of 'high trust', are more likely to gain the right to teleworking than their poorer brothers and sisters. Thus the technical fix might only serve to increase divisions of class within the sexes, as well as maintaining gender differences for some groups in society.

Scenario 3. Beyond Capitalism

The rules of work change. The working week is shortened to 32 hours. A decent national minimal weekly wage is established and those who choose to work less than that will be paid pro-rata. All those with caring responsibilities will be entitled to flexitime, and able to negotiate suitable arrangements with their managers. It is illegal for any organisation to require an employee to work more than 32 hours without special remit from the Department for Employment (to allow for conditions of special need and urgency), except in cases where they are covering for absent colleagues (on a voluntary basis). The long hours' culture vanished as people recognised the terrible toll it was having on their health and family lives. A volunteer system is well established where people who feel they have 'time on their hands' can contribute labour to community and public works. Such people, and others who are particularly concerned with making extra earnings, may set up their own businesses in arts and craft or small-scale service provision An improved state-run public transport system cuts down on travel to work time, and new businesses are located by local authorities on a rational basis so that everybody has a chance of finding work reasonably near home. Long-distance commuting is publicly frowned on, as people's values have shifted towards family and neighborhood.

A while ago, in a lively and provocative book, *Farewell to the Working Class* [25], the French sociologist Andre Gorz predicted that we were coming close to a social system in which the labour of the whole of an industrialised nation's populace would not be needed to carry out the total volume of socially necessary work. In effect, a lot of the population would become redundant, especially those from the manual sectors, where jobs would disappear because of automation and the use of robots. Nor would these jobs be replaced by new service-sector jobs, as computerised systems would also mean decreased employment in offices.

Gorz set out two possible scenarios for the future. One involved long-term employment becoming entrenched and increasing polarisation in society between the unemployed and employed. As Bauman [26] has suggested, the unemployed and marginalised would become enclosed in ghettoes, secluded from the rest of society, while those in employment flourished. Deploring this possibility, Gorz put forward another option: the necessary work should be shared out fairly between the populace meaning that everybody would only work for two or three days a week. The rest of the time could be devoted to family and leisure, voluntary and community work, or developing small businesses. Gorz implied that a rainbow alliance (greens, feminists, 'New Age' youngsters, the poor and excluded) might come together to campaign for such a future.

Alas, the present looks closer to Gorz's dystopian vision of social polarisation. With hindsight Gorz also drastically underestimated the extraordinary and elastic ability of capitalism to create new 'needs' and employ labour to fulfil them (from MP3s to personal trainers, from growers of pricey 'organic' foodstuffs to nail-painting bars). Yet my own post-capitalist and family-friendly scenario for the future is remarkably like Gorz's Utopia. At the heart of this vision is a major reallocation of time away from work.

To get there from where we are now, however, appears to run in the face of contemporary economic rationality, the logic that springs from capitalism. Thus we are told that in order to compete costs must be cut. Either people must work for lower pay, thus pressurising them into voluntary overtime to earn a 'living wage', or work processes will be intensified by forcing people to produce more for the same pay, thus driving up some people's working time. Popular resistance to the 'long hours' culture' that has resulted, especially in the UK, has proved limited. There is a certain type of middle-class masculinity that revels in showing its ability to thrive and compete in this environment. Cockburn [27] observed this 'heroic masculinity' in trade union leaders: it is also common among young professionals in areas such as law, IT and academia. Managers encourage it and those who embrace this culture are more likely to be rewarded and promoted [28]. All this disadvantages women whose domestic responsibilities bar them from competing, and is a disaster for women with children. As one University and Colleges Union activist concisely puts it, 'the bravado surrounding the long hours' culture needs to be displaced if women are to achieve genuine equality within the sector' [29]. But there are no signs of this happening. Nor is there any real challenge from national governments to the 'logic of capitalism' that pushes wages down and compels so many men to work overtime to support their families. Thus the prevailing pattern of father working full-time, mother adopting the role of secondary labour is perpetuated and continues to underpin the gendering of reproduction.

Scenario 4. Back to the Future

Disillusioned by having seen their mothers struggling with the triple burden of employment, domestic labour and emotional nurturance, a new generation of young women voluntarily decide to 'return to the home'. Traditional values about the importance of the mother-child bonding reassert themselves and men resume their traditional roles as breadwinners. Women do not give up career aspirations altogether, but seek jobs which will fit in with childcare or take long planned career breaks, returning when children reach school age. It is generally accepted that, though women's work may be valuable and economically helpful to the family, their commitment to motherhood means they will always be 'secondary workers'. With less labour market competition men's earnings rise and the idea of a 'family wage' at a reasonable level is reinstated.

Given the prevailing work relations described above, it is not surprising if women are viewing their futures as mothers and wage-workers with some concern. When I and colleagues at Bristol University interviewed young adults aged 20–35 in Bristol in the early 2000s we expected to find the same kind of egalitarian attitudes to marriage and parenting that Wetherell et al. had earlier uncovered, what I in earlier research had referred to as a 'climate of equality' [30]. In fact we discerned a more complex set of attitudes. While there was an acceptance of the ideals of gender equality, some young women told us that they felt pressurised into assuming the 'double burden' and guilty if their own inclination was to stay at home while children were pre-schoolers. As their mothers had rebelled against the role of stay-at-home mothers and housewives, demanding the right to 'go out to work', some of these young women were rebelling against the new norm of the working mother and the ideal of 'having it all.' In the words of two of our respondents:

> The guilt I can feel just being at home is quite tremendous. And we have an image of how we should be career women, do the housework and be a mother. But if I really talk to people who have children, they all love just to stay at home while the husband goes out to work.
>
> When you have children things do change and I think that my generation, where we were brought up to believe that we could do it, we could have it all, we could have everything and I think that's rubbish actually and I've really come full circle on that. And I think that you *can have it all*, but there's a big cost and it's usually the children that lose out.

Merryn Smith [31] has uncovered similar tensions among young professional women in Australia. The idea of 'choices' around career and motherhood seemed to have triggered deep anxieties about possibilities of self-fulfilment. The new female subjectivities she discerned among this generation were construed in terms of problems, of the dangers of missing out on either motherhood or career success: women seemed to be set up to fail.

Women of childbearing age struggle to reconcile these competing demands and there are some signs of a 'backlash' against 'having it all'. In a nightmare scenario for a feminist like myself, women might decide that equality is not worth the hassle and give up the struggle in favour of traditional reproductive arrangements, as is still the norm in many societies dominated by religious conservatism.

Scenario 5. More of the same . . . .

But not all women have the choice to return to the home. For many women around the world wage labour is a necessity to ensure the survival of themselves and their families. Even in the privileged West, the high costs of housing mean that both partners in a relationship may be compelled to hold jobs. The currently predominant solution in the UK is for men to work full-time, women part-time while the children are young; thus the majority of the labour of childcare and housework is carried out by the woman. I cannot see how, given the economic conditions discussed above, this is likely to change.

These established gender divisions appeared starkly in interviews carried out with women from Britain's Muslim communities [3]. Parents exercise tremendous pressures on the daughters to get married young and have children. Marriage and having children are almost synonomous: and women without husbands remain an object of pity. At the same time, parents are also keen for their children to achieve economic success, and young Muslim women are now participating in the European trend of girls outstripping their brothers in educational achievement. Many enter universities seeking degrees in prestigious areas such as law, medicine or accountancy and business. These young women are hard-working and determined; yet once they get married they will also be expected by their husbands to take responsibility for childcare and domestic management. Thus many women interviewed in this study explained how their jobs had to be fitted to school hours and family obligations:

> Due to marriage commitments I can't work in the private sector which means longer hours. I work in the voluntary sector because of flexible time and I can get home to continue child care.
>
> I am not going to give up my family commitments, as I said before, I don't actually see career as the be-all and end-all of everything [3].

Reproduction shapes the meaning of the lives of such women (and there are many non-Muslim women who share these priorities), but constrains their ability to achieve equality with men in the pressurised contemporary workplace.

Conclusions: The resillience of Gendering

Back where we started . . . .? Not quite. As Smith [31] points out, the female subject is open to constant reinvention. Feminine identities and roles have changed and in this era which (in the West) valorises choice and individuality, women are faced with a wider array of options. However, the theme of this chapter has been that prevailing economic parameters, of competitive global capitalism, work against the degendering of reproduction. Thus the processes of gendering described in the early part of my discussion are set to persist. Women are pressured into motherhood and that continues to limit their participation in the sphere of production. Moreover, the male subject appears less responsive to reinvention. Men all around the globe, East and West, assert masculinity through their ability to survive and thrive in the world of global competition.

But the dilemmas of femininity play out in different ways from those of previous generations. Sascha is a woman in her thirties who urged me to see the Louise Bourgeois exhibition. After a disturbed adolescence and involvement in a damaging hedonistic subculture, Sascha is getting her life together. She is desperate to become a mother, but in the past men have let her down. Currently she is completing a part-time degree to further her chances of career success, while working for a major multinational with an aggressive long hours' culture. Sascha has bought into this, working long hours and over weekends in order to pay for her study and her huge rent bill. Immensely capable and talented, she will undoubtedly succeed in her productive world. But, like Smith's Australian professionals, she will see her life as in some sense a failure if she does not also have children. How will she become a mother and how will this fit in to her workaholic lifestyle?

I cannot answer that question, but I do know that over and over again in the past decade I have been amazed and impressed by the strength, determination and ingenuity with which women confront the forces of gendering. Though I have emphasised the deeply embedded nature of gender divisions, the future may yet surprise us all.

References

1. Glucksmann, M. (1995). Why 'work'? Gender and the 'total social organisation of labour'. *Gender, Work and Organisation* 2(2) 63–75.
2. Bradley, H. (2007.) *Gender*. Cambridge: Polity.
3. Bradley, H., Healy, G., Forson, C., and Kaul, P. (2007).*Ethnic minority women and workplace cultures: what works and what doesn't*. Manchester: EOC.
4. Hollway, W. (1996). Masters and men in the transition from factory hands to sentimental workers. In D. Collinson and J. Hearn (eds) *Men as Managers, Managers as Men*. London: Sage.
5. Sample, I. (2008). Two coffees a day doubles risk of miscarriage, says US study. *Guardian* January 21st.
6. Hilpern, K. (2008). Unfit to be a mother? *Guardian,* January 15th.
7. Letherby, G. and Marchbanks, J (2007). *Introduction to Gender.* London: Pearson Longman.
8. Frost, J., Bradley, H., Levitas, R., Smith L., and Garcia, J. (2007). Loss of possibility: scientisation of death and the special case of early miscarriage. *Sociology of Health and Illness.* 29(7), 1–199.
9. Inhorn, M. and van Balen, F. (2002). *Infertility Around the Globe: new thinking on childlessness, gender and reproductive technologies*. Los Angeles: University of California Press.
10. Letherby, G. (1999). Other than mother and mothers as others: the experience of motherhood and non-motherhood in relation to 'infertility' and 'involuntary childlessness'.*Women's Studies International Forum* 22(3), 223–234.
11. Woollett, A. (1991) 'Having children: accounts of childless women and women with reproductive problems'. In A. Phoenix, A. Woollett and E. Lloyd (eds) *Motherhood: meanings, practices and ideologies*. London: Sage.
12. Franklin, S. (1997). *Embodied Progress. A cultural account of assisted conception*. London: Routledge.
13. Evans, F. (1985). 'Managers and labourers: women's attitudes to reproductive technology'. In W. Faulkner and E. Arnold (eds)*Smothered by Invention*. London: Pluto.
14. Boseley, S. (2008). Caesarean sections linked to higher risks of hysterectomies. *Guardian* January Ist.
15. Letherby, G. (2002). Claims and disclaimers; knowledge, reflexivity and representation in feminist research. socresonline.org.uk/6/4/letherby.html

16. Greer, G. (2008). Why do so many female artists put themselves in their work-often with no clothes on. *Guardian*, January 28th.
17. Firestone, S. (1971). *The Dialectic of Sex*. London: Jonathon Cape.
18. Piercey, M. (1985). *Woman on the Edge of Time*. London: Random House.
19. Barrett, M. and McIntosh, M. (1982). *The Anti-social Family*. London: Verso.
20. Wetherell, M., Stiven, H., and Potter, J. (1987). Unequal egalitarianism: a preliminary study of discourses concerning gender and employment opportunities. *British Journal of Social Psychology* 26, 59–71.
21. Wheelock, J. (1990). *Husbands at Home*. London: Routledge.
22. Dermott, E. (2008). *Intimate Fatherhood*. London: Routledge.
23. Dermott, E. (2005). Time and labour: fathers and childcare. *Sociological Review* 53(2), 89–103.
24. Badinter, E. (1982). *The Myth of Motherhood*. London: Souvenir Press.
25. Gorz, A. (1982).*Farewell to the Working Class*. London: Pluto.
26. Bauman, Z. (1998). *Work, Consumerism and the New Poor.* Buckingham: Open University Press.
27. Cockburn, C. (1991). *In the Way of Women*. London: Macmillan.
28. Casey, C. (1995). *Work, Self and Society: After Industrialism.* London: Routledge.
29. Feeney, A. (2008). *Election Address*. London: University and Colleges Union (UCU).
30. Bradley, H. (1999). *Gender and Power in the Workplace.* London: Palgrave.
31. Smith, M. (2008). The fertile subject: a psycho-social exploration of professional Femininities. PhD Cardiff University.

Part IV
Parenting

Chapter 10
Is Transferred Parental Responsibility Legitimately Enforceable?

Matti Häyry

Context

Two questions precede any considerations of reproductive genetics and gender. They are "Should people have children to begin with?" and "Who is responsible for children and why?" If it is irrational or immoral to produce offspring, as I have suggested elsewhere [1–3], reproductive genetics turns out to be a waste of time and gendered family roles obsolete. Tom Buller examines critically my views on this in Chapter 11. If, again, some people are more responsible for children than others, this should be made visible in the allocation of child-related burdens in society. In what follows, I offer some general comments regarding this second question for further discussion.

Questions

Parents have a special responsibility for the wellbeing of their children. If a young child accidentally hurts itself in the company of several adults, including its parents, it is understood that it is primarily the parents' responsibility to comfort the child. If the parents are absent, others may have obligations towards the child. But in their presence even the entitlement, let alone a duty, to interfere can be questioned. (Are you suggesting that I cannot take care of my own child?) Parental responsibility is closely intertwined with parental autonomy and authority.

Despite the recognised fact that parents have a special responsibility for their children, the state forces me, through taxation, to cater for the wellbeing of other people's children even when the parents are perfectly capable of doing it themselves. Money that I could use to buy goods and services of my choice is taken away from me and spent on healthcare services and educational packages for children whose parents could have afforded to pay for these. Why is this? Why am I, an individual

M. Häyry (✉)
School of Law, University of Manchester, Manchester, England, UK
e-mail: matti.hayry@manchester.ac.uk

F. Simonstein (ed.), *Reprogen-ethics and the Future of Gender*, International Library of Ethics, Law, and the New Medicine 43, DOI 10.1007/978-90-481-2475-6_11,

who has no children and does not intend to have any, made to provide for other people's children when these other people are around to do it themselves?

A standard response to this is that parents and the state have an arrangement. The state needs new citizens for its continued existence and parents are the ones who can satisfy this need. Parents, in their turn, need protection for themselves and for their families, and the state is the institution that can guarantee the required security. Consequently, an unwritten deal exists between the two parties. Parents produce new citizens for the state in exchange for protection; and the state refrains from interfering with family life as long as parents continue to deliver good, contributing members to the civil society and the state.

This response does not, however, explain why I should pay for the healthcare and education of other people's children when they could do it themselves. Parents and the state can well have an agreement, but since I am not a party to the agreement, it is far from obvious how I could be bound by it against my will. I do need protection for myself and it would probably be prudential for me to enter into some kind of a social contract that safeguards me against harm. But the continued existence of the state and the coming to being of future generations is not naturally in my interest. It can be, and has been, made to be in my interest by law. If I do not pay my taxes I will be fined and eventually imprisoned. But this is state coercion which, like all forms of coercion, requires a justification, or at least an explanation.

In the sections below, I will explore the nature, scope, and limits of the special responsibility parents have for their children. I will begin (in section "Shortcut Attempts") by showing that standard ethical theories do not adequately support the extension of parental duties to nonparents. I will then go on to examine (in section "Parental Responsibility") the moral basis of the "original" obligations people have towards their own children. After this I will show (in section "Breeders' Pact") how these obligations can be shared between all actual, prospective, and potential parents in what I call the "Breeders' Pact". Having expanded the scope of parental responsibility to its natural limits I will argue (in section "Reaching the Limit") that a further extension to nonparents is not justified. The implications of these findings will then be studied in conclusion (in section "Implications').

To avoid unnecessary misunderstandings, let me stress, at this point, two details. First, I am not arguing that childfree individuals have no responsibilities for children. Since we are all somebody's children, this would amount to saying that we are not responsible for each other at all. We are, and childfree individuals are; it is only that the responsibilities of parents are, if my arguments are sound, different. Secondly, I am not suggesting that the idea of a social contract is useless or detrimental. We live in communities and societies and need some agreement on how to live. But admitting that *some* agreement is required is not conceding that *any* particular pact has to be accepted without criticism. A contract between all people, parents and nonparents alike, could be considerably different from deals made by just one of these intrinsically competitive parties.

Shortcut Attempts

In the opening paragraphs above I asked the following question: "Why am I, an individual who has no children and does not intend to have any, made to provide for other people's children when these other people are around to do it themselves?" One theoretically possible answer to this question is that the requirement can be backed up by standard moral or political theories. Human nature, universal rationality, the greatest happiness, or political nous could dictate that the continued existence of the state and the coming to being of future generations are worthwhile goals, and that pursuing them justifies the infringements of my liberty. The aim of this section is to show, however, that general normative doctrines do *not* offer a satisfactory answer to the question, and that all these shortcut attempts to legitimise the extension of parental responsibilities fail in closer scrutiny.

A *teleological* approach, normally associated with Aristotelian and Aquinian teachings and Natural Law theory, takes as its starting point the perceived requirements of human nature. In order to reach our individual and collective goals, we need to stay alive, procreate, and search for the meaning of our existence. The natural laws that we can deduce from these necessities are an obligation to seek protection, nutrition, and shelter; a duty to produce offspring and to nurture it; and a responsibility to develop ourselves mentally and spiritually [4]. The second demand is thought to give us, as human beings, the shared task of having children and caring for them.

I have two problems with this justification. One is that the model attempts to be factual and normative simultaneously. It alleges that procreation *is* one of our main goals. Well, it is not one of mine. Faced with this objection, the model jumps directly from description to prescription, to claim that procreation *ought to be* one of my main goals. But how did this move happen? Surely the claim cannot be that I am unnatural or perverted if I happen to have slightly unusual aspirations. My other doubt concerns the implicit hierarchy of our goals in the view. With my money *not* spent on the healthcare and education of other people's children (when the parents can afford them) I could finance a personal research leave, during which I could further our understanding of the mental and spiritual aspects of the human existence. Why am I not allowed to do so? It would fit perfectly the frame of our natural obligations, given that some of our aims need not, for some hidden reason, be promoted above others.

A *deontological* view, best known in the context of Kantian ethics, finds the basis of morality in universal rules set by rational thinking. Arguably, an obligation to provide for other people's children could be founded on one of these. The test is to formulate a rule which has no internal contradictions and which could be accepted by any intelligent being. Can such a rule be found? I do not think so. In our case, it should say something like, "Always provide for other people's children, even when they can do it themselves!" This does not make much sense, either in connection with Immanuel Kant's ideas about parental responsibility, or within his more general theory.

According to Kant, children have a right to be cared for by their parents until they can take care of themselves [5]. I will come back to this facet of parental responsibility below (section "Parental Responsibility"). The point here is that children have a specific right to be nurtured *by their parents*, not by others. In more general terms, the proposed rule is equally unconvincing. The child support I am forced to provide enables parents to use their money on goods and services they prefer, instead of spending it on their progeny. Assuming that their preference satisfaction serves their life goals like my preference satisfaction would serve mine, the rule takes a broader form, namely, "Always trade your own goals, however important they are, for the goals of others when the occasion arises." Since Kant's fundamental principle of humanity [6] forbids us to use ourselves merely as a means to realise the ends of others, this is clearly a non-starter as an ethical imperative in his framework.

A *consequentialist*, or to be more precise, Utilitarian, justification for coerced child support could in theory be based on the happiness and wellbeing produced by the arrangement. If everybody's financial participation in the nurturing of some people's children accumulates the highest net benefit overall, then it is, despite the harm it inflicts on me in the form of lost opportunities, in a "positive classical utilitarian" analysis the best policy and ought to be assumed [7].

A broad difficulty with this type of defence is to show that this or that course of action will in fact generate the best results. How can we predict with adequate accuracy what the long-term impacts of the current policy, or its reversal, would be to all the parties concerned? And if different groups would be differently affected, which of them should carry the most weight in the assessment? The demonstration could be easier in "negative" variants of consequentialism, which emphasise the prevention of harm instead of the net optimum of benefits and disadvantages. If people simply try to avoid hurting others, complicated calculations are not necessarily needed. But this line of thinking is problematic in the context of parental rights. The best way to prevent all future harm to human beings would be to get rid of humankind altogether – no people, no pain. The entitlement of the state to promote reproduction cannot very well be based on this kind of philosophy. More "positive" forms of utilitarianism must be applied, and this will inevitably lead to disputes over the proper definition, measurement, and balancing of harms and benefits [8].

Another major issue in consequentialism, the role of possibly excluded minorities, was confronted explicitly in the theory of *justice as fairness* developed by John Rawls [9]. He argued, among other things, that inequalities in a just society should be arranged so that they benefit also those who are in the least advantaged position. His view expressed in this "difference principle" can be taken to mean that if parents or children are systematically worst off in our societies, then society could rightfully coerce me into financing their maintenance and endeavours. This may sound tempting to advocates of transferred parental responsibility, but I believe that the attempt to justify involuntary child support is, when the parents have the required means at their own disposal, at odds with the Rawlsian doctrine as a whole.

The theory introduced by Rawls contained three basic rules of justice – the primacy of basic liberties, fair equality of opportunity, and the difference principle – and a pecking order for these. Since the difference principle is ranked lowest, the other two have to be considered first. The first principle says that everyone is entitled to the widest range of basic liberties compatible with a similar range for everyone else. These include the right to own private property, an entitlement that the state would have to deny in our case. Parents of children are entitled to their property and I am entitled to mine. As long as they have voluntarily chosen their parenthood, fair equality of opportunity is observed and the difference principle does not apply. If the parents end up having less funds for their nonparental projects than I have for mine, this is a result of their own free decision making and does not justify a violation of my property rights. Note that in our specific case *children* will not be, by definition, among the least advantaged: I am asked to pay for healthcare and educational services *which the parents could afford*. This is not an instance of leaving children uncared for. Rather, it is an instance of parents refusing to make the sacrifices that come with their self-chosen role and then trying to compel me to share the burden through state enforcement.

Let me, then, turn my attention to the role of parents (in section "Parental Responsibility") and to the apparatus by which they can try to delegate it to others (in sections "Breeders' Pact" and "Reaching the Limit").

Parental Responsibility

The words "parent" and "child" are ambiguous. One can be a parent or a child generally speaking or the parent or child of particular individuals. With advances in reproductive technologies, one can be any of these in a genetic, gestational, legal, or social sense. And quite apart from family relations, children can be just young human beings. It is not always clear which meaning is intended or suggested in ethical discussions.

To clarify the situation in the ensuing sections, I will use the following definitions:

- *Parents* = the voluntary genetic and gestational mothers and fathers of particular individuals being discussed
- *Breeders* = actual, prospective, and potential voluntary parents of individuals; genetic, gestational, legal, and social; and other people who want to assume the responsibilities of parenthood
- *Childfree people* = nonparents and nonbreeders, including involuntary parents and involuntary breeders

Section "Parental Responsibility" will discuss the responsibilities of parents; section "Breeders' Pact" will extend most of these to all breeders; and section "Reaching the Limit" will argue that they should not be assigned to childfree people.

One intuitively appealing account of the foundation and nature of parental duties is given by Immanuel Kant in his treatise on jurisprudence, *The Science of Right*. Reproduction, according to Kant, is a natural and necessary human function, and women and men are, by virtue of their humanity, entitled to form conjugal unions with an intention of having children. Since entering the union and consummating it are free acts of the will, the spouses are morally responsible for the consequences of these acts – the most important of which is that persons are brought into the world without their consent. Kant states that these individuals, children, have "an original congenital right . . . to be reared by the care of their parents till they are capable of maintaining themselves; and this provision becomes immediately theirs by law; without any particular juridical act being required to determine it" [5]. He goes on to say that the act leading to the nonvoluntary emergence of the new people "attaches an obligation to the parents to make their children – as far as their power goes – contended with the condition thus acquired." So parents are responsible for their children's wellbeing, with all their strength, until the children can take care of themselves, because it is the parents' doing that they exist in the first place [10]. This sounds quite reasonable to me.

Some people, however, question this result, claiming that the acts that produced *their* children were not free or voluntary. Contraception was not properly available or failed on the occasion that led to pregnancy. Social pressure in the form of parental or peer group expectations, religion, ideology, education, or public policy made other alternatives impossible. Or one of the parents did not want to have children and only consented to it to save the relationship.

In all these cases it is, no doubt, true that people's reproductive choices have not been completely free. Perhaps this could be used as a mitigating factor when the relative responsibilities of parents, grandparents, families, friends, churches, politicians, teachers, and governmental offices are weighed. Two facts remain, however. The first is that *some* special responsibility can always be assigned to parents who have not been truly coerced into having (or wanting to have) children by force, threats, deception, or the absence of competence or information. The second is that even in the purely coercive cases *no* responsibility can be assigned to me. When the parents are not to blame, it stands to reason that the individuals and groups who created the pressure should assume the compensatory obligations, not innocent childfree bystanders.

Others question the content of parental responsibility when circumstances prevent people from discharging their duties towards their children. It is argued that if parents lack the economic or emotional means to nurture and school their young, society should step in and provide for these future citizens. After all, Kant's requirements are met when fathers and mothers have tried their best, regardless of the end result. If the children still need care, perhaps others should be involved.

At this point John Stuart Mill comes to the rescue by offering a useful restriction to Kant's conjugal entitlement. People who cannot look after children should not have them. As Mill wrote in *On Liberty*, "to bring a child into existence without a fair prospect of being able . . . to provide food for its body [and] instruction and training for its mind, is a moral crime, both against the unfortunate offspring and against society." Addressing specifically the issue of education, he continued that "if the parent does not fulfil this obligation, the State ought to see it fulfilled, at the charge, as far as possible, of the parent." Lamenting the situation of his own time, Mill observed that, on the contrary, "Instead of his being required to make any exertion or sacrifice for securing education to the child, it is left to his choice to accept it or not when it is provided gratis!" [11]. The overtone of the passages is that here parental authority has run amok while parental responsibility has been forgotten.

If Mill's sentiment is taken seriously, the right to have offspring postulated by Kant ought to be qualified and steps should be taken to ensure that people do not take procreation too lightly. Rather than telling individuals that they are free to have children whenever they feel like it, they could be given a threefold caution:

- If you cannot take care of it, you are advised not to have it.
- If you go on to have it but do not take care of it, the state will step in, in the first instance at your expense.
- If you do not have the means, the state will consider punishing you in an appropriate manner.

The idea would be to highlight parental responsibility already before children are brought into existence. Not only do people have duties towards their existing offspring; they may also have duties towards society (and possibly their nonexistent offspring but this is a metaphysically complicated notion) obliging them not to burden others with their unattended brood.

Breeders' Pact

The conclusions of sections "Shortcut Attempts" and "Parental Responsibility" may seem harsh, on at least two accounts. It could be argued that parents, as useful members of society, should not have lesser funds available for their nonparental projects than nonparents who do not contribute to the continued life of the nation. And it could be argued that since procreation is such a valuable enterprise, it should not be hindered even if in some cases joint efforts are needed to sustain the products of ill-considered propagation. Both objections rest on a contested premise, namely that human continuity is an inevitably good thing (I will return to this challengeable idea in section "Reaching the Limit" below), but they point the way to the next stages of my examination. The first of these is the need for a parental contract by which the burden of responsibility can be distributed more evenly within the community of breeders.

In their more lucid moments, actual, prospective, and potential parents must understand their responsibility for the lives that they have created or are about to create, imminently or in the more distant future. They must also understand that they will not always be around to shoulder the responsibility. Sudden death can take them away from their children; and illness, injury, financial hardship, marital problems, and changing social and political circumstances can make it difficult for them to secure their offspring's health and happiness. Their responsibility can, arguably, also be extended to future generations – to people who will be born after we have died. These people would not have come to existence without today's parents and their reproductive choices, nor would their hardships and suffering have occurred. Today's parents will not, however, be there to protect and comfort them when they need it.

Faced with these facts, parents must devise a scheme by which their dependent offspring will be looked after if and when they cannot do it themselves. Three strategies immediately offer themselves as candidates: direct coercion, commercial transactions, and mutual agreements. In the first option, direct coercion, parents bully others into taking care of their children by threats of violence or religious intimidation. This has worked in the past, in culturally contained communities in which hierarchies or traditions have had a strong hold of people's minds. As a modern solution, it has ethical and practical problems. Violence and intimidation are frowned upon, and in multicultural societies it would be difficult to predict how effective they would be, especially after the parents have died or otherwise become incapacitated. The second strategy, buying the services needed for one's offspring, sounds more promising, but it has at least two drawbacks. It is not a solution people with limited financial means could adopt, so it has an inegalitarian ring to it. It is also vulnerable to economic uncertainties and political upheavals like bankruptcies and revolutions. These issues make the third alternative seem the most palatable both in ethical and in pragmatic terms.

The third alternative is that breeders make an agreement (for the definition of a "breeder", see section "Parental Responsibility" above). They mutually promise that they will take care of each other's offspring. The general shape of the joint pledge could be something like the following.

> ***Breeders' Pact:***
> *When parents cannot guarantee the wellbeing of their children, then this shall be guaranteed by other actual, prospective, and potential voluntary parents and other people who want to assume the responsibilities of parenthood.*

The agreement could be made by limited groups of people like families, tribes, or communities. However, the wider the participation is, the better the coverage can be expected to be. Universal treaties can be difficult to reach except on a declaration level [12, 13], but national or regional consensus can be a realistic proposition.

The parties of the agreement can include people who have children (actual parents), people who are going to have children if all the biological and technological processes they have started are successful (prospective parents), people who intend to have children at some time (potential parents), and people who are not sure about their own plans but who have a positive attitude towards reproduction. Only voluntary participation can be binding. Individuals who have been coerced or duped into having children or into having propagation-friendly opinions by force, threats, deception, or lack of competence or information cannot even be held responsible for their own children, let alone other people's young.

The Breeders' Pact is easy to defend ethically. A mutual contract between parents and their voluntary associates is not susceptible to the weaknesses of the other two approaches (bullying and trade). It is not violent or coercive; it is open to all regardless of their financial fortunes; its good effects are predictable; and it is not immediately endangered by times of economic or political turbulence. If the consenting parties of the pact do not actually harm their children by trying to care for them (which sounds paradoxical but is a source of ongoing cultural and political debate), the arrangement seems ideal.

The nature and level of wellbeing guaranteed by the pack can vary historically, culturally, and socially. Women and men can have their own ideas about it, and so can people belonging to different generations and ethnic groups. Judging by contemporary bioethical controversies on parental autonomy and the type of children that people, according to competing theories, should be allowed to have, the main proposals seem to come from three schools of thought. The first emphasises unconditional love for one's offspring as a fundamental parental requirement and tells parents to ignore their children's physical characteristics, at least in the context of prenatal testing for adverse medical conditions or genetic mutations [14]. The second appeals to common sense and popular opinion and suggests that people should not have children whose health and social conditions would not enable them to live decently good and fulfilling lives [15–17]. The third approach insists that potential parents should always aim at the best possible lives for their children, especially in selecting them through preimplantation genetic diagnosis [18].

Responsible parenthood can be seen as a wider or as a narrower notion than the views expressed in discussions on genetic testing. These debates are prone to conflate various dimensions of the issue. Some focus on the choice of children before they are born and others on their treatment once they are more or less fully in existence. One group prefers to concentrate on physical wellbeing and another group on emotional welfare. However, bearing the risk of ambiguity in mind, the views can be used to outline nine theoretical ways of understanding the nature and scope of parental responsibility in the age of genetics and social change. These are presented schematically in Table 10.1.

The table's illustrative power is in that it lists both the views that people can actually embrace and opinions that their rivals would like them to have to make them easier targets of criticism. Box (a), for instance, represents a combination of ideas that champions of genetic testing would probably like their opponents to advocate. The view would claim that parents are not allowed to choose their children or their conditions, because children, whatever their physical or mental state, should be loved without any strings attached. An implication of this could be that it does not matter whether children are ill or not, and that if they have painful or debilitating conditions which could be routinely cured, parents should ignore the treatments and offer the children their unconditional love instead. It is highly unlikely that anyone would support a view like this, and more than probable that boxes (d) and (g) capture better the outlook opponents of genetic testing have. In a similar vein, box (j) could

Table 10.1 Some possible views on parental responsibility

The responsibility of parents for their children's		emotional wellbeing should be		
		unconditional	reasonable	conditional
physical	ignored	(a)	(b)	(c)
wellbeing	reasonable	(d)	(e)	(f)
should be	absolute	(g)	(h)	(j)

require parents to produce only the best children they can, with the proviso that if the result is less than perfect, parental obligations can be relaxed accordingly. Again, this box is likely to remain uninhabited, with the proponents of prenatal selection migrating towards positions (h) and (g).

The details of the Breeders' Pact are, as long as their agreement is free and uncoerced, theirs to decide. I will return to some of its characteristics below (in section "Implications"). Before that, let me explain why I believe that the requirements of the contract cannot be widened to childfree individuals.

Reaching the Limit

The foregoing sections have shown that taxing childfree people to provide for breeders' children needs a justification; that general ethical theories do not offer this justification; that voluntary parents are responsible for their children because they have caused them to exist; and that parents and their friends can obligate each other to take care of all children by a mutual contract. The question of this section is: Can the contract be rightfully extended to nonbreeders against their will? Or, is transferred parental responsibility legitimately enforceable? If the answer is yes, the justification needed for taxing childfree persons has been found.

The extension states that everybody should pay equally for the welfare of children, even when parents could afford it without help from others. This cannot be based on the fact that children have needs that must be satisfied. They do have needs and these must be satisfied, but in our context their parents can do this without involving me. The statement can, however, be based on the premise that we are all parents and all children are therefore our children. This is a false premise, but it is worth inspecting due to its explanatory value.

If everyone is or should be a breeder, then all people are or should be seen as our children, collectively speaking, and the shared parental responsibility defined by the Breeders' Pact requires us all to join our efforts in child care. The idea that everyone is or should be a breeder has been historically supported by the relative inadequacy of reproductive science and contraceptive techniques. When people cannot be sure about their fertility status and when birth control cannot be relied on, no one who engages in heterosexual practices can epistemically rule out the chance of becoming a parent. If this is accepted without protest, then universal breedership justifies the extension of parental obligations to all.

With advances in science and contraception, however, and with the social acceptance of terminations of pregnancy, it has ceased to be even remotely true that everyone is a breeder. There are quite a few people who are, like I am, knowingly and intentionally childfree. Involuntary reproduction remains a possibility, but that would have to happen without consent, and the responsibility for the offspring would rest with the people who instigated the process. What, then, about the claim that everyone *should be* a breeder? Perhaps the continuity of human life is such an important goal that we should all have a positive attitude towards procreation, if not by participating in it at least by sharing some of its burdens. This would lead to the same normative result as the factual line of argument – to the universal extension of parental duties.

Is the continuity of human life an inevitably good thing? Many people seem to think so, but it is difficult to find good grounds for the view. Is it good that humanity as a whole continues to exist? Not necessarily, because life is characterised by suffering. Why should we expose yet new generations to this, if the alternative is the peaceful demise of the race? Is it good that new individuals are brought into being? Not necessarily, and for similar reasons. They will inevitably suffer at some point in their lives, and if they are unlucky the suffering will be the predominant feature of their existence. In the light of these observations, one would think that the right question is, "Should *anyone* have children" rather than, "Should everyone have them?" [1–3, 19].

Since the Breeders' Pact can be thought of as a contract, and since contracts are in political theory often applied to parties against their will, it is useful to examine the conditions on which the extension could be legitimate. I will *not* consider this from the viewpoint of purely hypothetical agreements, though. Social contract doctrines typically postulate that every rational or reasonable person ought to grasp the overall advantage of succumbing to the "general will" of the majority or an enlightened minority, as life in political communities would otherwise be too unpredictable or hazardous to be tolerated. I agree that our life together requires rules, but a set of sensible rules does not inescapably include a universal duty to provide for children whose parents could easily do it themselves. Instead, I will approach the matter from an angle that is more sensitive to the plight of prudent dissenters.

In his defence of a "minimal state", Robert Nozick [20] argued that some people can be rightfully coerced into accepting the security mechanisms of a regionally dominant protective association, although they do not need the mechanisms and could in fact defend themselves better by their own private arrangements. The justification of the model is twofold. The end result is, for the members of the association, a spontaneous and uncoerced product of voluntary market transactions. And the ones restricted by the deal will be compensated by those who benefit from the existence of the association, also known as the minimal state.

Using comparable logic, would this argument make sense in the context of human reproduction? Like bodily integrity, the survival of the family line is purportedly important to most people. The Breeders' Pact could be construed as a voluntary agreement between parents and their friends, resulting in a state that secures the conditions of procreation and child care. The Pact could then be extended to bind childfree individuals, like Nozick's minimal state can be forced upon non-members of the protective association. Would this be the required justification for my duty to pay for the healthcare and education of other people's children?

No, not really. Being protected against violations of basic rights is in the interest of the non-members as well as the members of the Nozickian minimal state, whereas the survival of the family line is not in the interest of childfree persons. In one scenario the disagreement concerns only the methods of protection; in the other it concerns its aims. To make things worse, the requirement of compensation would make void the very norm the argument attempts to justify. Even if a breeders' state could be imposed on others, the ill effects of the imposition would have to be compensated to them. In other words, they would have to be paid money to cover their child-related taxes. Would it not be easier *not* to collect the taxes in the first place?

The line of thinking introduced here can, however, be used to one good purpose. One of the stock responses of parents to my view is to say, "Why would *our* children feed you in your old age, if you refuse to pay for their healthcare and education?" There are at least two answers to this question. Firstly, the way things are going, I am not convinced that anyone will feed me in my old age. The welfare states in which I live are efficient in collecting money from people who are currently working, but no one knows what is left of their pension functions by the time I reach retirement. Secondly, breeders' children should feed me in my old age, because I have involuntarily fed them while I have worked. Had I been allowed to keep the money that I have paid for their maintenance, I could have tried to secure my retirement by private investments. Either they or their parents are in possession of funds that belong to me. In all fairness, these should be returned to me.

Implications

In section "Breeders' Pact", I borrowed arguments on prenatal genetic testing to illustrate the different views people can have about the contents of the Breeders' Pact. The link is not coincidental. Many bioethical controversies can be seen as clashes between actual, prospective, and potential parents on the proper recognition and management of children's needs. Arguments about abortion and assisted reproductive technologies are obvious examples, but debates on patient autonomy, recreational drugs, and end-of-life decisions can easily acquire similar features. A possible explanation for this can be found in the fundamental moral principles that parents can easily resort to.

When parents and their ideological associates determine the content of their shared obligations towards children, the following universal formulation sounds plausible.

Breeders' Imperative:
Treat all children in the way that you would like your own children to be treated!

This rule has problems. Parents disagree with each other on the appropriate healthcare and education services that children should receive, and they have different views about the involvement of children in various types of decision making. But it is a rule that all those who endorse the Breeders' Pact can accept *for themselves*. There is an obligation to care for children; I know best how my children should be cared for; therefore I can assume the obligation to care for other people's children in a like manner. Treating my neighbours' children differently could imply that theirs is the right way, and this in its turn could mean that my original ideas were wrong. In obvious and nonthreatening cases, I can do to my neighbours' children what their parents tell me to do; when our opinions differ, I will try to get my own view recognised as the truth. I can acknowledge the rule as a good guideline for my conduct, although other people's interpretations of it can be disputed.

The relativity of moral judgements is, in general, openly admitted when it comes to the treatment of children. Parents have their beliefs, and so do teachers, medical professionals, social workers, state officials, nongovernmental organisations, scientific experts, and childfree individuals. The scope of parental autonomy and authority is in some areas challenged and confined, and the restrictions can vary according to the economic, social, and marital status of the parents. This reflects the fact that most people are breeders and they are engaged in an ongoing political battle for procreative supremacy. In this battle, medical, social, and scientific arguments are constantly employed to gain leverage over the parental intuitions of others.

The situation becomes murkier when child-related "parentalism" gives way to adult-related "paternalism". This occurs when the double meaning of the word "child" is allowed to shift the emphasis of the Breeders' Imperative. Two different formulations clarify the point.

Breeders' Imperative (child-related):

Treat all young people in the way that you would like your own children to be treated!

Breeders' Imperative (adult-related):

Treat all adults in the way that you would like your own children to be treated!

From the viewpoint of the Breeders' Pact, these sentences are almost equivalent. The deal is to care for everybody's children; and all people, including adults, are somebody's children. But this is where the underlying and for the most part unrecognised parental attitudes start generating scary normative views.

People do not usually like to neglect their offspring. If a young child comes to self-inflicted harm and if this could have been prevented by parental advice, precaution, or prohibitions, mothers and fathers customarily blame themselves for the damage. This is a suitable reaction and a healthy recognition of parental responsibility. It also explains why parents sometimes worry excessively about their offspring's choices and actions even after these choices and actions have become autonomous and independent of the breeders' possible mistakes.

The problem is that many people in public positions of authority carry this attitude to their dealings with their adult compatriots. They have internalised their original and contractual duties towards children so completely that they assume parental responsibilities for entire populations. If their fellow citizens come to self-inflicted harm as a result of their voluntarily chosen activities, these authorities imagine that this is their fault and start creating counselling services, protections, and prohibitions. People are denied medical services that they would like to purchase, notably experimental treatments that are deemed dangerous. The use of recreational drugs is limited or banned. Analgesics are prescribed sparingly in the fear of addiction and side effects even when competent patients are willing to take their chances with them. The process of dying is allowed to drag on by a reluctance to let people have a more active role in it.

But I am beginning to digress. The aim of this chapter was to show three things. People who have chosen to be parents are responsible for their children, because they are the reason for the children's existence. Voluntary parents and others who appreciate human propagation can and probably should endorse a contract by which all breeders become responsible for all children. This contract cannot be legitimately extended to bind childfree individuals. In addition to these three points, this concluding section has revealed some conceptual implications of the breeders' contract. Assuming the ethos of the deal can lead to the illusion that all people should be treated like children. Understanding that childfree persons are also a part of the wider social arrangement within which we live could mitigate the paternalistic threat presented by the current breeders' rule, where it exists. I have not specified how this could happen. I hope, however, that the points that I have presented here could pave the way to reflective discussion on the reasonable social and political obligations of parents and nonparents alike.

References

1. Häyry, Matti. 2004. A rational cure for pre-reproductive stress syndrome. *Journal of Medical Ethics* 30: 377–378.
2. Häyry, Matti. 2005. The rational cure for prereproductive stress syndrome revisited. *Journal of Medical Ethics* 31: 606–607.
3. Häyry, Matti. 2009. An analysis of some arguments for and against human reproduction. In *Arguments and Analysis in Bioethics*, Matti Häyry, Tuija Takala, Peter Herissone-Kelly, and Gardar Árnason (eds.) pp. 167–175. Amsterdam and New York: Rodopi.
4. Aquinas, Thomas. 1988. (orig. 1265–1272). *On Law, Morality and Politics*, W. P. Baumgarth and R. J. Regan (eds.) pp. 47–48. Indianapolis and Cambridge: Hackett Publishing Company.
5. Kant, Immanuel. 1887. (orig. 1796). *The Philosophy of Law: An Exposition of the Fundamental Principles of Jurisprudence as the Science of Right*. Edinburgh: T. and T. Clark. Trans. by W. Hastie. http://oll.libertyfund.org/?option=com_staticxt&staticfile=show.php%3Ftitle=359&chapter=55777&layout=html&Itemid=27. Accessed 5 February 2008.
6. Kant, Immanuel. 1994. (orig. 1797). *The Metaphysical Principles of Virtue*. Indianapolis and Cambridge: Hackett Publishing Company. In Immanuel Kant, *Ethical Philosophy* (second edition), 97. Trans. by James W. Ellington.
7. Häyry, Matti. 1994. *Liberal Utilitarianism and Applied Ethics*. London and New York: Routledge.
8. Häyry, Matti. 2007. Utilitarianism and bioethics. In *Principles of Health Care Ethics* (second edition), Richard Ashcroft, Angus Dawson, Heather Draper, and John McMillan (eds.) pp. 57–64. Chichester: John Wiley & Sons.
9. Rawls, John. 1972. *A Theory of Justice*. Oxford: Clarendon Press.
10. Shiffrin, Seana Valentine. 1999. Wrongful life, procreative responsibility, and the significance of harm. *Legal Theory* 5: 117–148.
11. Mill, John Stuart. 1869. (orig. 1859). *On Liberty* (fourth edition). London: Longman, Roberts & Green. http://www.bartleby.com/130/5.html. Accessed 6 February 2008.
12. *Declaration of the Rights of the Child*. 1959. Proclaimed by United Nations General Assembly resolution 1386 (XIV), 20 November.
13. *Convention on the Rights of the Child*. 1989. Adopted and opened for signature, ratification and accession by United Nations General Assembly resolution 44/25, 20 November.

14. Vehmas, Simo. 2002. Is it wrong to deliberately conceive or give birth to a child with mental retardation? *Journal of Medicine and Philosophy* 27: 47–63.
15. Purdy, Laura. 1995. Loving future people. In *Reproduction, Ethics and the Law*, J. Callahan (ed.) pp. 300–327. Bloomington, Indiana: Indiana University Press.
16. Freeman, Michael. 1997. *The Moral Status of Children: Essays on the Rights of Children*. The Hague: Kluwer Law International.
17. Herissone-Kelly, Peter. 2006. Procreative beneficence and the prospective parent. *Journal of Medical Ethics* 32: 166–169.
18. Savulescu Julian. 2001. Procreative beneficence: why we should select the best children. *Bioethics* 15: 413–426.
19. Häyry, Matti. 2004. If you must make babies, then at least make the best babies you can? *Human Fertility* 7: 105–112.
20. Nozick, Robert. 1974. *Anarchy, State, and Utopia*. New York: Basic Books.

Chapter 11
Reproduction, Responsibility and Rationality

Tom Buller

Introduction: Parental Rights and Responsibilities

In general terms, competent adults are deemed to have the right to reproduce. This right is generally understood as a negative right – a right of non-interference – derived from broader notions of autonomy and liberty. To say that someone has a right to reproduce is to say that, other things being equal, the person has the freedom to decide whether to reproduce or not, and other persons have correlative duties not to interfere. In this regard, the right to reproduce is akin to other negative rights, for example, the right to choose one's own religion, partner, or the method by which to educate one's children. Framed as a negative right, the right to reproduce serves to recognize and protect individual values and personal liberty against the interests of others or social utility. As we are familiar, this negative right is not universally respected, and some countries have enacted policies that permit such restrictions. Furthermore, we may believe, or come to believe in the future, that such restrictions are morally defensible.

The right to reproduce, however, is not a unitary notion, and neither is the right isolated nor absolute. As has been argued elsewhere, the notion of reproductive freedom should not be understood simply to be limited to the decision whether to reproduce or not; rather this freedom encompasses a number of decisions, such as choices regarding reproductive method and what kind of children to have [1]. Furthermore, as Tom Murray has argued, the right to reproduction conceived as "procreative liberty" may be too narrow a framework for thinking about parents and children [2]. Decisions to have children are rarely, if ever, made or defended purely in terms of negative rights; rather, they are made in the context of other interests, values, and obligations.

Additionally, the right to reproduce is not unrestricted. As John Robertson has argued the right to reproduce entails that the person violates no prima facie moral duty in deciding whether to reproduce or not; in other words, if S has a right to X then, other things being equal, S does not have a moral duty to refrain from

T. Buller (✉)
Department of Philosophy, University of Alaska Anchorage, Anchorage, AK, USA
e-mail: tombuller@uaa.alaska.edu

F. Simonstein (ed.), *Reprogen-ethics and the Future of Gender*, International Library of Ethics, Law, and the New Medicine 43, DOI 10.1007/978-90-481-2475-6_12,

doing X [3]. For example, S's duty to refrain from disclosing certain information means that S does not have the freedom to disclose it. However, this does not imply that the possession of a negative right is inconsistent with the existence of correlative duties and responsibilities. The right to free speech is not the right to say anything you like whenever you want to, and we thus hold that people are morally obliged to refrain from crying "fire" as a joke in a crowded theater; however, we do not conclude that such a moral restriction thereby invalidates the right to free speech. Similarly, in *Planned Parenthood v. Casey* the US Supreme Court upheld a woman's right to decide abortion whilst at the same time agreeing that a state could impose restrictions that did not impose an "undue burden" [4]. Rights as freedoms need not be absolute in order to be upheld. Moreover, even if one grants the coherence of a right to reproduce, this does not imply that every reproductive decision should be viewed in the same moral light.

Although we may grant that individuals (or couples) have a right to reproduce, the right to reproduce carries with it correlative duties for the prospective parents themselves. Minimally, these duties may be epistemic requiring that the individuals be sufficient informed. A recent example of this is the requirement by the State of Alaska that physicians provide information described in a state website to women seeking an abortion [5]. Alternatively, these duties may be more extensive and directly related towards behavior. Thus we expect parents to participate in appropriate prenatal care and to avoid behaviors that will predictably lead to birth defects. Societies have supported positive and negative social policies including mandating (and sometimes providing for) maternity leave, offering free prenatal care, issuing warnings about alcohol consumption and tobacco use by pregnant women [6], and other more controversial legal attempts to enforce "good behavior" among low-income pregnant women [7, 8]. Importantly, the enacting of such social policies and expectations is not regarded as the denial of procreative liberty, any more than denying someone the right to cry "fire" in a crowded theater is regarded as tantamount to the denial of free speech. Once the child is born the obligations of parents increase, as does the right of the state under *parens patriae* to paternalistically protect the interests of the child.

A "Minimally Satisfying Life"

The advancement of reproductive technologies has provided prospective information, and hence, usually, options, for those seeking to have children. What was once a matter of chance is now increasingly a matter of choice. These choices bring with them moral challenge in that the possibility of intervention (either active or passive) to prevent future harm or to produce future benefit opens up the possibility of responsibility. This responsibility, in turn, is framed by what we believe to be the boundaries of reproductive choice. Which ethical principles should frame these boundaries?

Perhaps the clearest, although not uncontroversial, cases that help set the boundaries of reproductive choice are those in which it is revealed that there is a considerable likelihood that the future child will have a life of "negative quality." For example, we can imagine the case of a child born with profound cognitive and physical disability whose future life can be predicted to be one of considerable suffering. In this case we might judge such a life to be "unfair" since it is not balanced by happiness, and whatever benefit the child's life brings, this benefit is unlikely to ever be experienced by the child itself. One might, of course, resist the notion that a life can have negative quality, for example, one might hold that all life is sacred and of infinite quality, or one might especially reject the idea that we can make such a judgment regarding a future life. In response it is worth stating the obvious that to hold that all life is sacred is neither to hold that all life has the same quality, nor that all lives should be maintained. It is also worth considering whether quality of life judgments in reproductive contexts are more epistemically challenging than quality of life determinations generally. Are there good epistemic reasons for maintaining that the life of the future child as described above will not be of negative quality?

If we accept that that some future lives can have negative quality, then following Laura Purdy, we might hold, therefore, that prospective parents have an obligation to provide a "minimally satisfying life" and hence should not reproduce if they are at high risk of transmitting a serious disease or genetic defect [10]. In a similar vein, Michael Parker has argued that prospective parents "have an obligation to ensure, in so far as this is possible, than any child they have has a reasonable chance of a good life" [11]. In this regard, we would be holding the right to "reproductive freedom" as constrained by an obligation of non-maleficence, as are other medico-moral decisions: if the life of the future child constitutes a "harm" (person or non-person affecting), then this harm should be avoided. Accordingly, if there is good reason to suppose that the life of the future child will be one of negative quality, then this would present us with *prima facie* reason for holding that prospective parents should avoid reproducing.

The notion of a "minimally satisfying life" leads to the formulation of the following principle of "procreative non-maleficence" (PN) according to which parents are to avoid having a child whose future life would be of negative quality. PN is consistent with both consequentialist and non-consequentialist moral thinking, and with a virtue ethics approach. Furthermore, PN appears *prima facie* justified on the grounds that the principle does not *necessitate* substantial restriction on procreative liberty, any more than a more general principle of non-maleficence does on overall human action. Certainly, one can imagine social policies that are consistent with PN that are too restrictive, for example, imposing criminal penalties for prospective parents who fail to undergo prenatal testing, but such restrictions are not necessary.

A different but related point can be brought out if we consider the relationship between PN and rationality, and the options and choices made possible by reproductive technologies such as preimplantation genetic diagnosis (PGD). PGD provides greater information and opportunity for prospective parents to identify factors that relate to the child's future quality of life. For the sake of argument, one can imagine that Smith undergoes PGD and has the choice between implanting a "normal" embryo or one which is revealed to have a severe abnormality, for example, a neural tube defect. According to PN, Smith ought to choose to the "normal" embryo, other things being equal. One way to gain a rough idea of the restrictive nature of a rule is to see whether people would choose to act in accordance with the rule, if the rule were absent. It is, of course, conceivable that Smith might wish to have a child with a serious neural tube defect; however, I think, that it is more likely that, given the choice, Smith would choose to implant the "normal" embryo. If this is correct, and prospective parents would make generally make reproductive decisions consistent with PN even if PN did not apply, then this provides some reason for claiming that PN does not overly restrict reproductive choice.

Thus we might initially conclude that PN is not too restrictive and hence a defensible reproductive principle. Nevertheless, it can be objected that PN is too restrictive since it would lead to the conclusion that it is always immoral and irrational to have children. An interesting and important version of this objection has been suggested by Matti Häyry [12].

Häyry's argument runs as follows. (1) A person acts rationally if, as a result of careful deliberation, he or she avoids choosing the action that leads to the worst outcome, consistent with the person's beliefs and values. For prospective parents the worst outcome is for the future child to have a life of negative value, hence prospective parents act rationally if they seek to avoid this outcome. (2) If prospective parents decide to have children there is a risk that the future child will have a life of negative value (either at birth or as the result of injury later in life), whereas if they decide not to have a child there is no risk that the future child will have a life of negative value. (3) Since the decision not to have children cannot logically be better or worse for the future child, the decision not to have a child has zero value. In conclusion, if parents act rationally so as to avoid the worst outcome they should decide not to have children.

H's argument can be summarized as follows (call this argument H1):

P1. It is rational to avoid the possible negative outcome when the alternative outcome has neutral or zero value.

P2. A decision not to have children has the value of zero in terms of the potential future individuals and their lives.

P3. A decision to have children has a possible negative value.

C. It is rational, therefore, to choose not to have children.

For many, the conclusion that it is always irrational and immoral to have children would be controversial, not least because universal adherence to the conclusion would have somewhat drastic consequences. Nevertheless, Häyry's argument is challenging and important because its conclusion appears to follow from a principle such as PN: if we have an obligation to avoid having a child whose future life could have negative value, and we hold that all future lives could have negative value, then the conclusion would appear to follow that we ought not to have children. As Häyry argues, the rule underlying his argument is the *maximin rule* (or strategy). According to the *maximin rule*, a rational agent should "adopt from the alternatives the worst outcome of which is superior to the worst outcome of the others" [13]. In other words, it is rational to choose the least-worse option.

I do not believe, Häyry's argument is successful, however, and thus we do not need to re-think PN. To see why, it is helpful to compare Häyry's argument H1 with the following one, H2:

P1*: It is rational to choose the possible positive outcome when the alternative outcome has neutral or zero value.

P2*: A decision not to have children has the value of zero in terms of the potential individuals and their lives.

P3*: A decision to have children has a possible positive value.

C*: It is rational, therefore, to choose to have children.

Do we have any grounds for preferring H1 to H2? *Prima facie*, there appears to be no reason why we should prefer one argument to the other and, more importantly, it seems difficult to imagine on what grounds Häyry could do so. For H2 is dependent on exactly the same *maximin rule* that we ought to avoid the least-worse option. The conclusion in H2 that it is rational and moral to have children, and in H1 that it is irrational and immoral to do so, both qualify on the grounds that they are the least-worse options. Moreover, P2* is identical to P2 and P1* and P3* seem no less defensible than their respective counterparts P1 and P3. If this analysis is correct, then we there would appear to be no reason for supporting H1 over H2. Hence we should not conclude that PN (or the *maximin rule*) leads to the conclusion that it is always irrational and immoral to have children.

A further conclusion to draw is that one cannot prohibit (or support) every instance of reproduction purely on *a priori* grounds. In broad terms, an action is rational if it is the product of some reflective process and is broadly consistent with the person's goals, interests and values. What would it mean to say that each and every decision to reproduce is irrational and immoral? It would mean that no matter what the circumstances, and no matter what a person's interests, goals, and values, it is impossible that such a decision should turn out for the good. This seems to be a claim that could never be epistemically justified. Rational decision-making should relate to facts of the matter in regard to probability, risk, and other actions: if Jones wants to travel from London to Paris but hates flying, and he has the choice between driving or flying, then he *should* choose to drive. Probabilities, preferences, and facts matter.

The Best Life or a Good Life

If the above criticism of Häyry's argument is correct, then one can justifiably conclude that the decision to reproduce is not always irrational and immoral. For this and other reasons stated above I believe that one can conclude that PN is not too demanding to serve as a defensible principle of reproductive responsibility. Perhaps, though, PN is not demanding enough? The discussion so far has focused on what may arguably be regarded as the least controversial cases warranting a limit to reproductive freedom, namely those in which the future life is predicted to be of negative quality. If, however, the future life is expected to be at least "minimally satisfying," then does this mean that, other things being equal, further limitations on reproductive freedom are unwarranted? For the sake of argument, imagine that a couple has undergone preimplantation genetic diagnosis (PGD) and have the choice to implant one of three embryos, A, B, or C. On the basis of the genetic information, A is predicted to have a life of average quality, B is predicted to have at least a minimally satisfying life, and C is predicted to have a less than minimally satisfying life. For example, PGD might detect A to be genetically "normal," B to have a predisposition to asthma, and C to have a genetic abnormality that always leads to profound physical and cognitive impairment. According to the discussion above, PN would dictate that that the couple should choose not to implant C but the principle is neutral regarding the choice between A and B. Since both of these lives are of at least reasonable quality the principle does not appear relevant and we would have no justifiable reason, therefore, to constrain the right to reproductive freedom. Prospective parents are obliged to avoid selecting future children whose life would be harmful, but they have no obligation beyond this, for example, to choose to select the "better" oven the "best" child.

In an important article Julian Savulescu has argued for a principle of "Procreative Beneficence" and the view that prospective parents do have an obligation to select future children with the "best possible life."

Couples . . . should select the child, of the possible children that they could have, who is expected to have the best life, or at least as good a life as the others, based on the relevant, available information [14].

Savulescu provides the following case, the *Simple Case of Selection for Disease Genes,* in support of the principle of procreative beneficence (PB).

A couple is having IVF in an attempt to have a child. It produces two embryos. A battery of tests for common diseases is performed. Embryo A has no abnormalities on the tests performed. Embryo B has no abnormalities on the tests performed except its genetic profile reveals that is has a predisposition to developing asthma. Which embryo should be implanted? [14].

According to Savulescu, "on pain of irrationality" Embryo A should be implanted since Embryo B has "nothing to be said in its favour." On all aspects but one A and B are identical and in the one that differentiates them this difference gives an advantage to A. In such a case it would seem to be difficult to justify why prospective parents would not choose to implant A. We might disagree with Savulescu that asthma is disadvantageous or question the description of the options; however, if we accept the case as presented it seems difficult to defend the choice to implant Embryo B rather than Embryo A. Of course, a prospective couple might prefer B on the grounds that they believe that asthma correlates with high IQ or creativity, and perhaps they are correct. But this would be to change the example because it would now be the case that something could be said in favor of B and, therefore, this would make the decision to choose B over A rationally defensible.

However, although one might agree choosing embryo B would be irrational, this conclusion is not sufficient to show that such a decision would be morally questionable. What is needed is a reason to think that prospective parents are morally required to seek to implant the embryo with the best possible life, rather than avoiding selecting those whose quality of life would be minimal. A plausible candidate is the common-garden or folk view that parents, prospective or actual, are expected to act in the "best interest of their children." Importantly, when such a claim is made we do not generally regard it as imposing an impossible burden or committing parents to reaching an impossible goal; for "acting in the best interest" can be understood in a moderate way as saying that parents should choose the best option, other things being equal. And the qualifying phrase means exactly that: if all other things are equal then parents should choose the best option; in other words, we would have a moral expectation that prospective parents should choose A over B since this is exactly how the case is described.

Savulescu's argument has prompted considerable discussion [15]. In a recent article Michael Parker has criticized Savulescu's account on the grounds that the principle of procreative beneficence is "undetermining, paradoxical, self-defeating, and overly individualistic" [16]. Here I will focus merely on the first of these criticisms. According to Parker, Savulescu's principle of procreative beneficence is undetermining because plausible accounts of the best life are too broad and complex to be reduced to facts about the genetic constitution of the embryo. As Parker says:

... any coherent use of the principle of procreative beneficence in ranking possible lives would unavoidably involve ranking the characteristics of, say, embryos, in relation to a cluster of complex, rich and interdependent moral concepts. This is not possible for two reasons. The first of these arises from the very fact that complex concepts, such as those of the good life, the best life, and human flourishing, are not reducible to simple elements or constituent parts which might be identified through the testing of embryos ... [The second reason is that] the diversity of preferences for, and beliefs about, the relative importance of such elements, combined with the variety of their possible interactions means that it would not, even in theory, be possible to identify the rational choice with respect to any particular feature of an embryo or possible child [17].

To highlight the difference between the accounts of Savulescu and Parker it is helpful to consider embryos A and B above. On Savulescu's account the prospective parents should implant embryo A on the grounds that it has no abnormalities whereas embryo B carries the predisposition for asthma. For Parker, however, the genetic predisposition for asthma as revealed by PGD would tell us little about whether the future child will have a good (or the best) life, since the future quality of life is dependent upon a myriad of factors and not merely the embryo's genetic status. Furthermore, this myriad of factors would make it impossible to make a rational and informed decision to rank, and therefore, to choose one embryo over the other. Since we have no good reason for preferring embryo A to embryo B it would be irrational to select one embryo in preference to the other; in other words, prospective parents should be morally neutral between A and B.

As I understand it, Parker's argument contains two core claims that are uncontroversial: firstly, that quality of life is rarely determinable exhaustively by genetic factors; and secondly, a decision to select an embryo on the basis of genetic factors is, therefore, epistemically difficult. The contexts of quality of life preimplantation reproductive decisions are customarily, though not always, opaque – the lack of a clear relationship between genetic disposition and quality of life generally prevents ranking embryos according to their genetic status. What I believe is more challenging is why acceptance of these claims presents a particular problem for Savulescu's account. The underlying thought seems to be that whereas we are able to clearly identify cases in which the future life will have negative quality, we are less able to identify those in which the future life will have positive quality. Part of the reason for this difference might be that our notions of the "good life" are likely to incorporate higher-order preferences, intentions and be contextual, and identifying, and agreeing upon, what counts as the "best life" are impossible tasks. More importantly, in the former (negative quality) but not the latter (positive quality), the genetic status of the embryo is a stronger predictor of future quality of life since environmental or social factors are less relevant. Once we have reached the threshold of a minimally satisfying life the future quality of life is contingent, unpredictable and influenced by a myriad of social factors.

If this understanding is correct, then the implicit objection is that Savulescu's account commits one to an unlikely degree of genetic determinism. There are, I believe, a number of responses to this line of objection. First, although Savulescu's focus of discussion is on the selection of embryo in cases of PGD, this focus can be broadened to be more consistent with the notion of "acting in the best interest" described above. For example, imagine that B's parents have the option to move to a environment that will be better for B's asthma, a move that will have no net reduction in B's or the family's welfare. In such a case we can imagine that the overall harm in not doing so might be significant, and this might lead us to conclude that the family ought to move according to a principle of non-maleficence, rather than beneficence. However, even if this point is granted, it would still extend the range of parental obligation beyond the requirement of a "reasonable chance of a good life."

Second, the argument can be made that the accounts of Parker and Savulescu are fundamentally similar for they both argue that prospective parents should select whichever is the better of A or B. For Parker, all possible embryos are divided into two groups: those that have a reasonable chance of a good life, and those that do not. Presumably, Parker would argue that those in the first group should be ranked higher than those in the second group and thus, other things being equal, those in the first group should be implanted at the expense of those in the latter. If we define the "best" life (or the "better" life) as those lives with a "reasonable chance of a good life" then the embryos that would be ranked highest on Parker's account would be identical to those ranked highest on Savulescu's account. The implication here being that the accounts of Parker and Savulescu differ as to the criteria for ranking embryos, but they agree as to the rational and moral obligation to prefer the higher ranked over the lower ranked.

Reproductive Choice

The above suggests that the principle of procreative beneficence may be no more morally demanding than competing principles. If one understands this principle as a comparative one that obliges prospective parents to choose the "best" embryo from available alternatives, and this obligation is an example of the more general one for parents to act in the best interest of their children, then it is not evident that such a principle of procreative beneficence would undermine reproductive freedom. Certainly, if "best" is allowed to be defined in purely subjective terms then PB would have no force or relevance, and if "best" is determined by social rules that took precedence over subjective preference, then this freedom would be threatened; however, there is no clear reason why either of these options should be chosen.

If one is prepared to support the obligation of procreative beneficence and the more general obligation then this will have implications for sex selection. In the following chapter in this book David Heyd argues that sex selection by prospective parents for non medical reasons is defensible on the basis of parental autonomy (although it is not a specific right) [18]. Briefly, his argument is that our objection to sex selection can be understood not as an objection to sex selection *per se* but to artificial cases and furthermore, that objections on the basis of demographic danger and discrimination are unfounded. On the basis of the discussion in this paper it seems that one could make the stronger claim that not only is sex selection for non-medical reasons defensible, but that it is morally required. To many this may well appear to be an extraordinarily absurd claim; however, I think that it can be defended.

As Heyd mentions in his article, and as we are familiar, a person's gender can play a significant role in a person's quality of life. In many, perhaps all, countries to a variety of degrees, male children are regarded as preferable to female children; men have greater autonomy, power, influence and payment for their work than women. If this description is accepted, then it would appear straightforward that prospective parents ought to select male children rather than female. Moreover, is there good reason for thinking that sex selection requires an essentially different moral analysis than other types of selection? It seems plausible to contend that if we believe that prospective parents are under a general obligation to act in their (future) child's best interest, and we are prepared to defend embryo selection on this basis, then there does not appear to be any *prima facie* reason why sex selection should be regarded as an exceptional case.

Conclusion

In this chapter I have attempted to elucidate the general principle that informs reproductive choice, and support the principle of procreative beneficence as a viable candidate. In broad terms I believe that prospective parents, like actual parents, are morally required to act in the best interests of their children. This means that if prospective parents have the choice between selecting one of two embryos they are morally required to select the best, other things being equal. Finally, I wish to reiterate the earlier point that this view does not need to entail that parents have any less authority in determining what is in their children's best interest than rival accounts.

References

1. Buchanan A., D. Brock, N. Daniels, and D. Wikler. 2000. *From Chance to Choice: Genetic and Justice*. Cambridge: Cambridge University Press, 209–213.
2. Murray, T.H. 2002. What are Families For? Getting to an Ethics of Reproductive Technology. *Hastings Center Report* 32(3): 41–45.
3. Robertson, J. 1994. *Children of Choice: Freedom and the New Reproductive Technologies*. Princeton: Princeton University Press.
4. *Planned Parenthood v. Casey*. 1992. 505 U.S. 883.
5. Section 18.05.032 of the current Alaska Statutes. The website is located at: http://www.hss.state.ak.us/dph/wcfh/informedconsent/default.htm
6. Section 04.21.065 of the current Alaska Statutes.
7. *Ferguson v. City of Charleston*. 2001. 532 U.S. 67, 72.
8. Capron, A. Punishing Mothers. *Hastings Center Report*. 1998. 28(1): 31–34.
9. Steinbock, B. and R. McClamrock. 1994. When is Birth Unfair to the Child? *Hastings Center Report* 24(6): 15–21.
10. Purdy, L.M. 1996. Genetics and Reproductive Risk: Can Having Children be Immoral? *Reproducing Persons: Issues in Feminist Bioethics*. Ithaca: Cornell University Press.
11. Parker, M. 2007. The Best Possible Child. *Journal of Medical Ethics* 33(5): 279–283.
12. Häyry, M. 2004. If You Must Make Babies, Then at Least Make the Best Babies You Can? *Human Fertility* 7(2): 105–112.
13. Rawls, J. 1971. *A Theory of Justice*. Cambridge: Harvard University Press, p. 153.
14. Savulescu, J. 2001. Procreative Beneficence: Why we Should Select the Best Children. *Bioethics* 15(6): 413–426.
15. Parker, M. and Stoller S.E. 2007, 2008. Why we are not Morally Required to Select the Best Children: A Response to Savulescu *Bioethics* 22(7): 364–369.
16. Parker, M. 2007. The Best Possible Child. Journal of Medical Ethics 33(5): 279–283, 281.
17. Ibid.
18. Heyd, D. 2009. "Male *or* Female we Will Create Them: The Ethics of Sex Selection for Non-Medical Reasons," in F. Simonstein *Reprogene-Ethics and the Future of Gender*. Dordrecht: Springer.

Chapter 12
Male *or* Female We Will Create Them: The Ethics of Sex Selection for Non-medical Reasons

David Heyd

Introduction

It is sometimes instructive to start from the beginning. On the sixth day of creation, God created man, "in the image of God He created *him*; male and female He created *them*" (Gen. 1: 27). God's choice is to create man in the generic sense (which reflects God's own a-sexual nature). The split of male and female, that is the move from "him" to "them", is only derivative, though necessary, since only by such a split can humans pro-create themselves, thereby reflecting God's unlimited creative power [1]. But then there, is of course, the other version of the creation story. God first created Adam, "man" in a specifically male form, and only upon realizing that "it is not for man to be alone" he decided to make "a fitting helper for him" (Gen. 2: 18). If, as I read the biblical text, the image of God in which human beings are made amounts to the power of procreation, then the two versions of the creation of "man" suggest two alternatives for approaching the problem of sex selection. According to the first, control over human reproduction extends only to the act of procreation, while the sex of the future child is left to luck or to the work of nature. According to the second version of the story, we may in principle have the power of choice between having a male or a female offspring.

Advances in modern medicine have given human beings safer and more reliable methods of birth control, that is the exercise of choice over the number of children we have. But now, with the development of ultra-sound technology, amniocentesis, and most recently pre-implantation genetic diagnosis (PGD), we have gained also the ability to choose the sex of the child. We are getting closer to the power to choose at least some elements in the genetic character of future children, and hence their identity, even beyond their gender. To most people, some control over the number of children and the timing of their conception looks morally innocuous, while designing the genetic identity of future children seems morally abhorrent. In between lies the question of sex selection: may we act on the basis of what has become an easily accessible information about the sex of a future child?

D. Heyd (✉)
Department of Philosophy, Hebrew University, Jerusalem, Israel
e-mail: david.heyd@huji.ac.il

F. Simonstein (ed.), *Reprogen-ethics and the Future of Gender*, International Library of Ethics, Law, and the New Medicine 43, DOI 10.1007/978-90-481-2475-6_13,

I would like to argue that from a philosophical perspective, sex selection is a "non-issue". Unlike abortion, sex selection does not involve the killing of an allegedly unique human being; unlike human cloning, it does not raise questions of the nature of reproduction; unlike surrogate motherhood, sex selection does not force us to revise the notion of parenthood and the institution of the family; and most conspicuously, unlike genetic engineering, it does not give rise to a revision in our deep conception of human identity and the long-term nature of the human genome. This does not mean that there are no moral considerations that should be taken into account when we form an opinion about practices of sex selection; and indeed these considerations will constitute the substance of this paper. My only argument is that the extensive treatment of the subject in the last few years, in both public and professional circles, has often had the tendency to over-dramatization, as if sex selection involves radical challenges to traditional conceptions of humanity and reproduction.

Fears and Negative Connotations

There is a strong emotional element underlying the current rhetoric about sex selection. The word "selection" itself is marred by horrifying connotations relating to the infamous separation of people on the railway platform in a concentration camp. The term "selection" is also unfortunate in its being associated with the preference of items that are considered objectively superior, like tomatoes in the market. Accordingly, the "select" or "selected" are those items, or human beings, that enjoy a special standing due to their high quality. But for a sober ethical discussion of the determination of the sex of future people, a more neutral term could serve us better. "Choice", in contradistinction to "selection" implies only a preference and not necessarily an objective evaluation of quality or a general norm. But since the term "sex selection" has become so widespread, I am going to use it in the rest of the discussion, though with a value-neutral sense in mind.

The emotionally charged debate about sex selection should be understood on the background of the general reluctance in society to "play God", to intervene in natural processes, to assume control over what traditionally has been left to fate or luck. Although we often have definite wishes and hopes regarding our offspring, we are not enthusiastic about attaining control over their realization. We regard many aspects of reproduction as processes that "happen" rather than chosen and hence tend to a traditional policy of non-interference. The law is typically a conservative system of norms, which is often slow to adapt to new options in reproductive technologies. Thus, most European countries today as well as many other countries in the world prohibit by law the practice of sex selection for non-medical reasons, and are cautious and restrictive about selecting the sex of an embryo on medical grounds [2].

The most common apprehension about sex selection has to do with the confusion of the choice of sex with the intervention in the human genome. This is a powerful fallacy that must be dispelled before we engage in the more rational arguments against sex selection. The source of the mistake is easy to detect: sex is indeed one of the properties that constitute the identity of human individuals as are other genetic qualities currently studied by genetic science. However, it is obvious that by choosing the sex of a child we do *not* interfere or manipulate the genome in any way, especially when we do so for *non*-medical reasons. Hence, the common concern with a "slippery slope" leading from sex selection to genetic engineering has no basis in reality and the discussion of the two issues should better be completely separated in both professional and public discussion. Sex selection has nothing to do with eugenics and does not lead to it.

Another related confusion that plays a major role in the debate on sex selection is the allusion to Nazi practices. The alleged slippery slope from sex selection or some forms of genetic screening and engineering to Nazi experiments and mass murder is illusory. The inhumane and cruel treatment of human beings practiced by the Nazis was propelled by a racist hatred to Jews, Slavs, Gypsies and other non-Aryan people rather than by a eugenic plan for the betterment of the *human* species. Eugenics preceded Nazi ideology, on the one hand, and Nazi ideology developed mostly independently of eugenic theory, on the other. The wish to choose the sex of one's child is a personal or a cultural preference, which is not related to a eugenic or racist ideal of a world consisting of a single sex.

Methods of Sex Selection

There is nothing new in the human attempt to decide the sex of future children. The reasons for preferring one sex (usually the male) over the other are of many kinds: economic (work force, the cost of dowries); prestige and inheritance; religious; and psychological (family balancing). Numerous methods have been tried in the course of history, most of which proved ineffective and lacking scientific basis. The Talmud reports one such folk theory:

> R. Isaac b. Ammi stated "If the woman is first to emit the semen she bears a male child and if the male is first to do it she bears a female child" [3].

Some commentators note the considerate attitude to the woman's sexual welfare is implied in this counsel, namely, if the man wishes to have a son, he should do his best to let his wife reach an orgasm first. But the scientific basis of this advice is highly dubious. The Talmud itself, in another place in the same Tractate, concedes that the only way to enhance the chances of having a male child is prayer!

Diets, timing of sexual relations, or the sexual positions of the partners have all been suggested as methods for determining the sex of the child. None have proven reliable (or, as the joke goes, they have a 50% chance of success). But it should be noted in the context of our discussion that none of the home-made methods has ever given rise to an opposition on ethical grounds.

Modern science has provided us with a variety of scientifically-based methods of determining the sex of the future child. These can be ordered according to the stages in the reproductive process. A pre-conceptive method involves sperm sorting, followed by artificial insemination. In a later stage, in vitro fertilization (IVF) might be used to create a number of fertilized ova out of which those of the desired sex are selected for implantation in the uterus. The most effective way of doing so is by pre-implantation genetic diagnosis (PGD) whereby a single cell is removed from an eight-cell "pre-embryo" and tested for a whole gamut of potential genetic disorders. The sex of the embryo can be easily discovered in this procedure. Ultra-sound and amniocentesis are tests undertaken in yet a later stage of gestation and in some societies are used as methods of sex selection. Finally, and unfortunately, sex selection is often practiced after birth, usually in the form of female infanticide or abandonment. This last "method" is of course not a new one and is not related to technological advances. One may also want to add to this list an even later phenomenon, even if not a systematic method, that expresses the preference of one sex to the other, namely the better chances of survival of children of the desired sex who are given better care, nutrition and protection.

This hierarchy of methods of sex selection is of ethical significance. The lower one is on the ladder, the less problematic is the practice from a moral point of view. We have mentioned already that "folk methods" are not considered morally problematic. The interesting question is whether the lack of opposition to them has to do with their inefficiency or rather with the fact that they are natural. Think of an imaginary DIY set, sold in a pharmacy, which could help couples decide the sex of their offspring: would that be considered ethically wrong? Legally prohibited? It seems that even if private methods of controlling the sex of the future child were to become effective and reliable, society would not want to interfere in the reproductive freedom of the parents anymore than it does in the case of family planning (i.e. decisions about the number of children and their spacing).

So it seems that the ethical reservations about sex selection have to do with the *artificiality* of the methods used for its realization and the need for external (medical) assistance. Sperm sorting is indeed a highly artificial procedure that interferes with the natural process of reproduction in a significant way. Its ethical advantage lies, however, in avoiding the controversial issue of abortion, since it precedes the existence of an embryo. No "person", even in the most minimal sense, maintained for instance by Catholics, is harmed by this method. And of course there is no risk to the mother either. It is true that the method is not very reliable nowadays, but technically it could become in the future quite easy to apply by couples without the need of any medical assistance.

IVF is different in morally relevant ways. The selection method is performed on fertilized eggs or pre-embryos, some of which are implanted in the uterus, others either destroyed or frozen. Being an invasive procedure, IVF also involves risk to the woman. Unlike "natural" methods and sperm sorting, which harm no one, IVF affects both existing pre-embryos and the mother. Medical reasons usually justify this procedure, but it leaves open the question whether IVF may be performed for *non*-medical reasons. In order to identify the sex of the future child a genetic test (PGD) must follow the IVF. Now, assuming that IVF is performed as a matter of medical necessity (e.g. to overcome a fertility problem), should PGD be permitted with the sole purpose of selecting the sex of the child? The more conservative view is that sex selection is justified only if its purpose is of a medical nature, typically to prevent sex-linked diseases. In other words, "social" sex selection should never be permitted. The more permissive view holds that if IVF has been performed for a medical purpose, there is no reason to prohibit the further PGD test, even if it is carried out just for the sake of selecting the sex of the implanted embryos. A middle approach between the two views is that sex selection could be performed only if there are medical indications for the PGD test, that is to say, selecting sex is legitimate only as a "side benefit" of a procedure that is independently justified.

The conservative view is based either on the slight yet existing risk to the embryo connected with PGD, or on the intrinsic wrongness of sex selection for non-medical reasons. The permissive view maintains that there is nothing wrong *per se* in sex selection and that the risks of PGD (unlike IVF) are minimal and hence do not override the wish of the parents to choose the sex of their child (or specifically decide which of the fertilized eggs, which after all belong to them, should be implanted). The middle way wavers between the two views, respecting the wish of the parents but doing so only when it does not require taking a specific intentional action.

The conservative view does not seem compelling because the risk in PGD itself is only marginal in a way which does not override either medical reasons such as the prevention of sex linked hereditary disease or non-medical reasons like parental autonomy. Furthermore, sex selection as such is not morally wrong, since, as we shall see in the following sections, it does neither undermine the demographic future of society, nor constitute a stigmatization of women. The middle way, namely applying sex selection only in cases in which PGD is performed for other medical (genetic) indications, does not seem consistent. For if PGD in itself does not involve risk to either the mother or the future child, why should it not be done for gender selection? And if gender selection is illegitimate, why allow it when it is a side benefit? Would it not be a discrimination against women who were not treated for a medical problem and hence are deprived of the right to choose?

The only argument which carries some weight is that doctors should not engage in practices that are not purely medical (such as sex selection for "social" reasons). In other words, the moral prohibition on social sex selection is related to the ethics of medical practice and the professional role of physicians, which must be limited to medical treatment. The scarce resources in medicine and the precious time of doctors should not be devoted to non-medical purposes. Unlike harmful genes, gender in itself is not a pathology and hence should not be considered a factor in medical treatment [4]. In response, it should be noted that physicians are nowadays directly involved in a whole spectrum of practices and procedures that are not "medical" in the traditional sense, be it cosmetic surgery or, closer to our concern, contraception and family planning (including abortions for personal reasons). We do not want physicians to become agents of society in implementing demographic or religious values.

Methods or practices of sex selection in later stages of the development of a human being are undoubtedly more problematic or even repugnant. Many believe that abortion, following ultra-sound or amniocentesis tests, is morally controversial and usually justified, if at all, only by serious medical or psychological considerations. A preference for one sex rather than another is hardly such a consideration. Infanticide or the neglect of children of a certain sex is a widespread practice in some societies, but of course can never be justified. But if infanticide is so morally abhorrent, should not abortions such as those performed in some clinics in India be considered the lesser evil? And if such abortions are also considered morally wrong, should we not prefer PGD, which is not an abortion, and does not harm the woman?

From the discussion so far we may conclude that if we look at the hierarchy of the stages of human development we can make the following judgments. Sex selection as part of the sexual act itself is morally neutral, and so is the case of sperm sorting. Performing IVF just for the sake of selecting sex is morally problematic and better avoided because of the risk and pain to the woman. But once IVF is performed for justified medical reasons, further PGD procedure for selecting a male or a female embryo is morally permissible. Abortion for choosing the sex of a child is morally problematic due to both the status of the fetus and the harm to the mother. Infanticide of all kinds is obviously wrong. Now, since folk methods and sperm sorting are far from reliable, we are (at least for the time being) left with PGD as the only permissible and effective means of sex selection.

However, our analysis so far has taken into account only the interests and rights of the mother, the embryo/future child [5] and possibly the professional duties of the doctor. But the most common arguments against sex selection focus on its impact on society. There are two lines such of opposition to sex selection: the demographic and the feminist. We will discuss them in turn.

Demographic Imbalance: Real or Imagined?

The principal opposition to the practice of sex selection is based on the fear that it might lead to a dangerous breach of the gender balance in future society. According to this approach, even if sex selection is morally permissible from the point of view of both the future child and the interests and rights of the mother, it should be prohibited on social grounds or with view of the harm to unidentified future people who are going to suffer from the scarcity of partners. The most conspicuous empirical evidence for the rationality of such fears comes from the widespread practice of sex selection in some countries, particularly China and India [6]. Without entering into the widely documented literature on the causes of the clear preference of males to females in these societies, we can say that the prohibition on more than one child per family in China has given rise to female abortions and infanticide so as to guarantee a male offspring. In India, culturally based biases and economic considerations relating to the disastrously expensive costs of dowries have led to the mass appeal to ultra-sound tests followed by abortions of female fetuses. Abortions and more recently PGD tests have become more attractive alternatives to female infanticide that seems to have been practiced for a long time in India.

Without detracting from the repugnance of female infanticide or abortions for the sole purpose of sex selection, it should be noted that they have been practiced for a long time before the introduction of modern technologies. PGD for sex selection is not only a morally superior means but at least at the moment has only an infinitesimal impact on the demographic balance. PGD is an expensive test and requires an invasive procedure (IVF) which yields only low chances of live birth (20% on average). Very few women would undergo this expensive and painful process just for having a boy rather than a girl. Immigration, for instance, has a much deeper influence on the demographic balance than medically assisted sex selection, and so do changes in child mortality that affect the two sexes differentially. Again, it seems that our moral reservation about sex selection is closely related to the active intervention in natural processes rather than to the demographic imbalance as such.

The concern about the impact of sex selection on the demographic balance should be seen in the wider perspective of large-scale demographic changes in human history. In her classic study, the Harvard social historian Marcia Guttentag analyses patterns of gender distribution in various societies in human history and their impact on the status of women in those societies [7]. Although it appears that there were fairly sharp differences in the ratio of men and women in certain societies (in both directions), none of them collapsed for that reason. Imbalances tend to correct themselves and societies adapt to them through institutional changes in social and legal practices. After all, even in a male-oriented society, men need women for the creation of more men![8]

Furthermore, the problem of preference for males over females is conspicuous only in certain societies, mainly in East Asia. In contrast, studies show that in most western countries there is no marked bias towards either of the sexes. Parents usually prefer a "balanced" family, consisting of children of both sexes [9]. Even in Jewish Orthodox society, in which there is a special premium on a first-born male, once a boy is born, the sex ratio in the rest of the children is of no significant concern [10].

So it seems that much of the demographic scare associated with sex selection is more imaginary than real. But even if sex selection by PGD or other modern technological means proved to be a demographic risk, some countermeasures could be devised for neutralizing it. Thus, for example, every institution practising PGD would be required to balance the annual number of implantations of female and male embryos. This might happen as a natural process (if indeed people prefer balanced families rather than children of a particular sex, as seems to be the case in the developed world). Or, alternatively, a market mechanism could be introduced, in which each couple who wants to have a boy would have to be "set off" by a couple who wants to have a girl. This arrangement of a Noah Ark, into which only "couples of couples" are allowed, may be a fantastical thought experiment, but it is a philosophical challenge to the demographic objectors to sex selection. One can also restrict the gender choice to parents who have already one child or have an imbalanced number of children [11].

Furthermore, the demographic worry could be easily allayed by a bold move that would condition the request for implantation of fertilized ova of a particular sex by the parents' consent to donate all the other eggs to another woman. This would not only solve the problem of demographic imbalance but also reduce the scarcity of fertilized eggs, which are in high demand by infertile couples. This proposition is admittedly problematic, since in many countries the donation of fertilized ova to other couples is legally prohibited. But this prohibition should anyway be reconsidered for reasons other than the facilitation of sex selection.

The last comment on the demographic issue concerns the relation between sex selection and population control. Be it with the purpose of family balancing or with the purpose of having more children of a particular sex (usually boys), the prohibition on sex selection contributes to population growth. Couples do not stop procreating as long as they do not have a child of a certain sex. If we want to curb population growth, particularly in third-world countries, we should allow early-stage sex selection (either by sperm sorting or, in the future, by cheap and risk-free IVF and PGD procedures) as a means for satisfying parental wishes in a more responsible way.

The Fear of Stigmatization: The Feminist Argument

Since in most cultures, if there is a marked social preference for one sex it is for the male sex, feminists have naturally associated sex selection with patriarchy, sexist biases, and female subjugation. It is undeniable that female infanticide or abortion for non-medical reasons reflect a general discriminatory attitude to women and that the social status of women in societies in which these practices are widespread is low. The feminist argument is that sex selection, even by means other than infanticide or abortion such as PGD or sperm selection, should be prohibited, since it reflects an andro-centric view and humiliates women. That is to say, even if there is no medical harm to the mother or to the embryo, the very idea of choosing the sex of children is wrong.

Marcia Guttentag, whom we mentioned above, studied in much detail the social implications of imbalanced sex ratios in various societies throughout human history. Her principal hypothesis suggests that the smaller the relative number of women to men, the higher is the status and prestige of women in society. In such societies, monogamy is stronger, although women are expected to fulfill their domestic roles. When the relative number of women increases and becomes larger than that of men (as was the rapid process in American society between the 1940s and the1970s), women lose their traditional prestige, they become more of sex objects, the number of divorces rises, one-parent families become more prevalent, and more women suffer from depression and suicide. However, in those circumstances more women become ambitious and career minded [12].

So the first response to the feminist argument against sex selection is that the decrease in the number of women might lead exactly to the increase in their social status, at least in some respects. But independently of this empirical argument, one can add a second response, namely that personal preference for male *or* female children does not stigmatize either sex and should not be considered a manifestation of a sexist bias. Personal choice should be clearly distinguished from a systematic social norm, policy, or institutional preference. Personal preference *might* express a gender bias, but by no means must do so. The typical wish parents have for a boy after having three girls does not indicate a male chauvinist attitude. Nor does such a preference manifest a prejudice in other circumstances, such as the wish of a single mother to have a girl rather than a boy, or even the wish of a father to have a son who will carry on his name. None of these preferences expresses in a general way the superiority of one gender to the other.

It is indeed true that some cases of gender preference do indicate a prejudicial and discriminatory attitude. But here we get to the third response to the feminist opposition to sex selection, which might be referred to as putting the cart before the horses. Sex selection does not *create* an anti-women bias; it is a *manifestation* of it. In other words, the way to tackle the problem of sex selection in India is to fight against the deep causes underlying it, to change the social structures and norms that make the birth of girls economically burdensome, like the dowry system. Girls in India are not unwanted as such; they are too expensive for many families to raise. And this can be changed by social reform. Criminalizing sex selection does not address the structural causes that give rise to it [13]. Thus, the current egalitarian attitude to the birth of boys and girls in the West is not the outcome of a prohibition on female infanticide or sex selection but a manifestation of the more advanced status of women in society, economically, politically and culturally. In a strange way, the success of the feminist general struggle for gender equality renders the issue of sex selection irrelevant. Nevertheless, feminists can still argue that the absence of a law prohibiting sex selection for non-medical reasons is a strong symbolic indication that society is indifferent to androcentric attitudes among its citizens. In other words, the law should be used in this case as an essential tool for initiating a change in social norms about gender since it is more effective in creating such a change than other forms of public education. This is definitely a strong argument which is empirically compelling, but it must be balanced with the general caution in liberal societies with regards to the use of legislation for moral education.

Conclusion: The Liberal View of Sex Selection

Reproductive rights have become entrenched in both legal and moral discourse in the modern world. Marriage and procreation are two dimensions in which people express their deepest preferences and life plans. Although the decision regarding the existence, number and identity of children has typically far-reaching consequences for society, it is protected in liberal normative systems as a private matter in which the law representing public interest should interfere only in extreme cases. Is sex selection one of the extreme cases? We saw that it lies somewhere between the choice of having children (including their number and spacing), on the one hand, and choice of their genetic profile (particularly by genetic screening, engineering and cloning), on the other. I argued that sex selection is closer to the former than to the latter, since it does not interfere with the human genome and does not introduce into it irreversible changes. In that respect, choosing to have a child and choosing its sex are no different in a morally relevant way. In gender choice human beings simply extend their power of creating "man" in the generic, conjunctive sense of Genesis 1 to creating Adam or Eve in the gender specific, disjunctive sense of Genesis 2. If the first kind of (pro)creation is an exercise of the image of God, so must the second be.

Liberalism does not mean complete neutrality of the state. Even in matters pertaining to procreation, the state has a legitimate interest, both in the paternalistic protection of its citizens and in promoting social interests of future generations. Thus, it may encourage or discourage certain practices, or create incentives and disincentives by means of taxation and the distribution of social services [14]. But these are *indirect* means which apply on the social level, and they should be well distinguished from direct intervention in the life of individuals and their choices. It is therefore consistent from a liberal point of view to establish institutional rules and conditions, such as those suggested above, so as to prevent unbalanced sex ratios, even when no harm to the mother or the embryo/child is involved. For example, it seems that even in a socialized system of health services, the state should not be required to fund sex selection for non-medical reasons.

So far most of the chapter was devoted to the refutation of the arguments against sex selection, such as the theological, the demographic, the feminist and the medico-ethical. The discussion of the issue in a negative method is natural to liberalism, which is more concerned with explaining why a practice is not wrong and should be permitted than with positively supporting it. However, the positive reasons for sex selection should be explicitly stated too. The exercise of personal autonomy of parents in choosing the gender of their child does not consist merely of a capricious preference. The wish to have a gender-mixed family is reasonable and unrelated to any bias or prejudice [15]. On the assumption that the parents prefer mixed families, would not two mixed families be superior to two unmixed families? Then there often are serious psychological reasons for having a child of a particular sex. One example is the hypothesis that a girl is easier to raise by a single mother than a boy. Another example comes from the only Israeli case so far that has reached a semi-legal discussion in which doctors asked for permission to select female pre-embryos in an IVF procedure for religious reasons [16]. A simple analogy can support this liberal approach. In adopting a child, the preference of the parents for a child of a particular gender is taken as both legitimate and reasonable. It is usually not regarded as consisting of any prejudice or gender bias.

By defending a liberal argument about the permissibility of non-medically indicated sex selection I wish by no means to argue that gender choice is a moral or a political *right*. Having a child of a particular sex is not a claim parents have against doctors or the state. It is not a protected interest that must be actively advanced by society. Society may view the choice itself as morally unworthy and not deserving of assistance (financial or other). It may even want to educate its citizens not to opt for such a practice and ask doctors to discourage it. The liberal argument advanced here made only the more modest claim of the *permissibility* of sex selection.

The philosophical analysis of the problem of sex selection in this chapter exposed the implicit distinctions that underlie but often confuse our judgment: the determination of the sex of the embryo before its creation (sperm sorting) vs. its determination after conception (PGD); the legitimacy of folk techniques vs. that of professional medical involvement; personal choice vs. public policy; the opposition to sex selection as such (for theological or feminist reasons) vs. the opposition to the practice on the basis of its consequences (demographic or social); performing PGD with the intention of selecting sex vs. getting the opportunity to do so as a side benefit. Dissembling the reasons for rejecting sex selection may reveal whether our opposition is absolute or conditioned, sweeping or circumstantial, principled or pragmatic. The liberal approach adopted in this paper views the practice of sex selection as morally permissible in principle, qualifying it only on pragmatic and circumstantial considerations.

It seems that with the future advances in reproductive technologies, we will gain the godly power to create male *or* female. Society is in the beginning slow to adapt to such radical changes in human control over natural processes, but it turns out that after a certain period of time it rapidly incorporates them to its benefit. Such was the case with organ transplants, with IVF, with genetic screening and with surrogate motherhood. Twenty years ago, with the introduction of ultra-sound test, people were horrified at the very idea of *knowing in advance* the gender of the fetus and many parents demanded that this information not be divulged to them. I can only guess that in ten years' time sex selection will become a widespread practice and would be treated as routine and morally innocuous.

References

1. I have developed the interpretation of procreation as the very image of God in my, "Divine Creation and Human Procreation: Reflections on Genesis in the Light of *Genesis*," in Nicholas Fotion and Jan C. Heller, eds., *Contingent Future Persons: On the Ethics of Deciding Who Will Live, or Not, in the Future* (Dordrecht: Kluwer Academic Press, 1997), pp. 57–70.
2. Section 14 in the Oviedo Convention of the Council of Europe prohibits sex selection for non-medical reasons. In the U.S. there is no regulation of the practice, but social sex selection is not "encouraged" (Recommendations of The Ethics Committee of the American Society of Reproductive Medicine in *Fertility and Sterility* 72 (1999): 599. In India, abortion for sex selection is a criminal offence. When this article went to the press, the British Human Fertilisation and Embryology Agency (HFEA) published its report "Sex Selection: Options for Regulation" in which the overwhelming majority of both individuals and institutions consulted expressed opposition to the practice of sex selection for non-medical reasons. See http://www.hfea.gov.uk/AboutHFEA/Consultations. The evidence for such a strong public objection to sex selection should be taken into account in the formation of regulatory policies, but it is not directly relevant to the critical normative discussion undertaken in this paper. See also John McMillan, "Sex Selection in the United Kingdom," *Hastings Center Report* 32 (2002): 28–31. In Israel there has been so far only one case which was given a semi-legal attention (see below). There is not much discussion in Jewish rabbinical literature about the subject, but it is agreed that sex selection should not be practised just as a matter of personal preference. The main reason is that this goes against nature. But rabbis have no problem with sex selection by PGD when its purpose is of a medical nature. See Richard V. Grazi and Joel B. Wolowelsky, "Preimplantation Sex Selection and Genetic Screening in Contemporary Jewish Law and Ethics," *Journal of Assisted Reproduction* 9 (1992): 318–322.
3. *Babylonian Talmud*, "Tractate Niddah," p. 26a (London: Soncino Press, 1959).
4. This argument was suggested to me by Asa Kasher. It is interesting to note that the attitude of physicians to sex selection is deeply mixed, even confused. In an ESHRE PGD Consortium study, 15 centers expressed opposition to the practice and only 4 a favourable attitude (with 2 abstaining). See *Human Reproduction* 17 (2002): 244–245. Yet, it seems that there is a sufficient number of doctors who are willing to help parents choose the sex of their children in medically assisted pregnancies.

5. By the interests of the child I mean only those of an *actual* child, i.e. a child who was born and is killed if it belongs to an "unwanted" sex. One cannot ascribe to the child either the interest in being born a male or a female (or not being born male *or* female) since possible children do not have interests or rights. On this complex issue, often associated with "wrongful life" cases, see David Heyd, *Genethics: Moral Issues in the Creation of People* (Berkeley: University of California Press, 1992), Chapter 1.
6. For the Indian scene, see Kusum, "The Use of Pre-Natal Diagnostic Techniques for Sex Selection: The Indian Scene," *Bioethics* 7 (1993): 149–165. Kusum argues that sex selection in India should not be permitted due to the particular current economic and cultural circumstances. But see a letter to the editor by Dr. Aniruddha Malpani, a famous doctor in a Bombay clinic, who openly defends the right of women to choose the gender of their child on liberal grounds. *Human Reproduction* 17 (2002): 517.
7. Marcia Guttentag and Paul F. Secord, *Too Many Women?* (Beverly Hills: Sage, 1983).
8. But then, as Alan Buchanan noticed, a problem of free rider is created: although all parents have an interest in leaving a world of balanced sex ratios to their children, they might have overriding personal interests in having a male child. If everybody had acted according to their personal interests, a Tragedy of the Commons would be created; if only few did so, they would be free riders. See, A. Buchanan et al. eds.,*From Chance to Choice: Genetics and Justice* (Cambridge: Cambridge University Press, 2000) pp. 184–186.
9. Thus, studies reveal that the number of parents who decide to have a *third* child after having two children of the same sex is significantly larger than the number of parents whose first two children are of different sexes. And even more important, there is no significant difference in the wish to have a third child between those who have two boys and those who have two girls.
10. The commandment to be fruitful and multiply was sometimes interpreted as being fulfilled by having at least one son and one daughter. Of course, there is value in having as many children as one can have beyond one boy and one girl, but then their sex becomes unimportant.
11. This is suggested in S. J. Fasouliotis and J. G. Schenker, "Preimplantation Genetic Diagnosis: Principles and Ethics," *Human Reproduction* 13 (1998): 2243.
12. Guttentag, Chapter 1. Guttentag, like others, has noticed the relative large number of men in Jewish Orthodox society. The general high regard for women in that society corroborates her hypothesis (including the values of monogamy and the domesticity of women). The causes of this unequal sex ratio in traditional Jewish communities is striking and has led to various attempts to explain it, mainly in terms of the rate and timing of sexual relations which are strictly regulated by rules of purity and the husband's duty to sexually satisfy his wife. Since male babies have a smaller chance to survive childhood, the generally lower child mortality in Jewish society may also be a cause of this surplus of men. See Guttentag, Chapter 4.
13. Cf. B. M. Dickens, "Can Sex Selection Be Ethically Tolerated?," *Journal of Medical Ethics* 28 (2002): 335–336; and Bonnie Steinbock, "Sex Selection: Not Obviously Wrong," *Hastings Center Report* 32 (2002): 23–28. Steinbock correctly argues that the prohibition of sex selection will not reduce sexist attitudes in society, but adds that the existence of a discriminatory society might be a reason for parents not to bring a girl into the world. This argument is susceptible to the same criticism as are wrongful life claims, since it ascribes rights and interests to possible people.
14. Consider the analogical case of IVF treatment after a certain age. Society has definite interests in discouraging it, for both the financial cost involved and the risk of the birth of handicapped children who would be a burden on society. Most of us believe, however, that criminalizing these parental choices would violate liberal principles and the privacy of individuals.
15. For a strong defense of the liberal approach to sex selection, based on the distinction between the coerced and the un-coerced exercise of the mother's choice in reproductive matters, see Mary Anne Warren, "Sex Selection: Individual Choice or Cultural Coercion?," in Helge Kuhse and Peter Singer, eds., *Bioethics* (Oxford: Blackwell, 1999) pp. 137–142.

16. An Orthodox couple who needed sperm donation due to the infertility of the husband was concerned that due to the special religious status of Cohen (priest) of the husband, the whole community would become aware of the fact that he was not the "real" father once the child reached the age of 13 and the Bar Mitzvah ceremony. Only the biological sons of a Cohen become Cohanim with the distinct duty of making a special priestly blessing in the synagogue. A female child would spare the parents this embarrassment since girls are not called upon to make a blessing or read the Torah in this public ceremony. In 2002 the Israeli Ministry of Health granted permission to select the female pre-embryos but did so in a purely ad hoc manner. Nevertheless, it attracted some criticism and a fear of a slippery slope of other demands for sex selection for non-medical reasons.

Part V
Non-gendered Futures

Chapter 13
Artificial Reproductive Technologies and the Advent of the Artificial Womb

Frida Simonstein

Introduction

The idea of an artificial womb and ectogenesis – a child brought to term outside a biological womb – has appeared in various thought experiments; [1–3] several movies have also shown some frightening scenarios based on this idea. And while some medical sources avoid discussion on this topic [4], other sources choose to address the advent of the artificial womb as a matter of fact [5].

The first reaction of many people when confronted with the idea of ectogenesis is revulsion and/or fear. However, some babies, those prematurely born, spend less time in a woman's womb. Therefore, and perhaps against our basic intuitions, the question of why it is required to be sometime in a woman's womb (nine months; six; five? two days?) cannot be automatically overruled.

Presently there is still a huge gap between the first stages of gestation (by IVF) and the 22nd week of gestation (inside the womb). For many, this gap seems an insurmountable barrier. Yet, the artificial womb might become a reality sooner than we may think: developments in neonatal care, gynecology, embryology, computer science and the human genome project are converging to this end.

This chapter presents the scientific, technological and legal developments towards the advent of the artificial womb and ectogenesis. It is not the purpose of this chapter to engage with a throughout analysis of the ethical consequences but mainly, to delineate the uncharted waters towards ectogenesis. While the artificial womb is the tool and ectogenesis the process, in this paper I will address both terminologies interchangeably.

Neonatal Care

Faye Bland was the first baby who was born six months prematurely, weighing only 470 grams and survived. This happened back in 1975; yet the survival of this baby was considered at the time a medical 'accident' since he was not supposed

F. Simonstein (✉)
Department of Health Systems Management, Yezreel Valley College, Yezreel Valley, Israel
e-mail: fridas@yvc.ac.il

F. Simonstein (ed.), *Reprogen-ethics and the Future of Gender*, International Library of Ethics, Law, and the New Medicine 43, DOI 10.1007/978-90-481-2475-6_14,

to survive. By the early seventies, doctor's efforts were focused on trying to save babies weighing around 1,500 grams; and those weighing less than 1,000 grams were allowed to die because nothing much could be done for them. Moreover, ten years earlier nobody would have tried to put in neonatal care a baby as premature as Faye Bland was [6]. Today, there is nothing extraordinary about the survival of a baby weighing 470 grams; [7] furthermore, the record of the tiniest baby who has survived is 300 grams; her name is Or Iluz a baby girl who was born in Israel in 2000 weighing at birth only 300 grams; by 2004 she was four year old [8].

Neonatal care has now expanded from caring for neonates who had need only a few hours, or days, of special care after delivery, into an ever-expanding frontier of prolonged, aggressive treatment for smaller and smaller infants. Yet these infants are dependent on external life support; as they move from the natural womb of their mothers to 'technological wombs' of a Neonatal Intensive Care Unit (NICU). Saving the lives of these babies might be considered an obligatory act in some parts of the world; [9] however, the outcomes of these savior acts are not always encouraging; because the improved survival of early and/or extremely low birth weight (ELBWs) and gestational age infant may be accompanied by increased risks of chronic medical problems and neuro-developmental disabilities [10]. In order to improve outcomes neonatalogists think that an 'artificial placenta' can be used in the future. As Jack Sills from the Intensive Care Nursery, University of California said at the *American Re-HealthCare Symposium 2000:*

> [i]t would be ideal to continue the in uterus environment, keeping the premature infant in a warm water bath (free of infection) attached to its artificial placenta. Under ideal physiologic conditions the premature would grow and develop normally, free of injury [11].

Sills was not talking completely 'out of the blue': Thomas Schaffer at Temple University has developed an amniotic fluid, which helps to mimic the way a fetus breath in the womb. And in Japan, Yoshinori Kuwabara, a professor of obstetrics at Juntendo University reported that he has actually created an artificial womb, using an acrylic tank filled with a fluid similar to the amniotic fluid at body temperature. This artificial womb has successfully developed goat fetuses to term. The animal fetuses have lied submerged in this 'tank womb' attached by the umbilical cord to a machine which acted as a placenta bringing oxygen and nutrients to the fetuses and acting also as a dialysis machine, cleaning the fetuses' blood [12].

Will this Line of Treatment Persist?

There is some disagreement about what should be done with ELBWs and prematurely born infants, because of the huge costs of NICUs; which continue rising due to novel improvements. Developments, however, paradoxically, make it possible to save the lives of more ELWBs and premature babies [13]. There are already databases tracking outcomes for babies 401–500 grams; This weight group of micro-preemie is considered 'experimental' since mortality in this group approaches

90% [14]. However, these databases suggest that this line of pushing back the barriers of neonatal survival is unlikely to perish, because sporadic successes make neonatalogists ask questions such as 'how small is too small' and 'how much is too much' for providing care to premies at the threshold of viability [15–18].

This approach is controversial because of the morbidity, the suffering and the huge costs. However, it is not unreasonable to suppose that if neonatal morbidity could be avoided, outcries against the huge costs of NICUs may disappear. Moreover, there is a societal conditioning in some societies towards the approval to the efforts for saving the life of a newborn – no matter its condition; and regardless of the outcomes. Hence, there is a generous support for neonatal care generally; and research in this earliest frontier between life and death is integral of neonatal care. Furthermore, since an artificial womb – or 'placenta' – may improve outcomes for ELWBs, it is quite improbable that there could be a serious obstacle to this development.

New Developments in Gynecology and IVF

While research activity on embryos at the end line of gestation is supported generously because it may save the lives of tiny humans, the beginning of gestation is seen very differently; research at this stage of development is highly controversial; and in most countries research on embryos beyond day 14 after fertilization is banned. Canada has explicitly prohibited any 'experimentation which may lead to ectogenesis' [19]. However, the report for artificial reproduction regulations equally exhorts for better results in IVF cycles [20]. Which cannot be pursued without further research on the process of embryo implantation in the womb; which is, clearly, research on ectogenesis. It follows that the proposal for ARTs regulations in Canada is incoherent.

Fertility clinics are under pressure to publish their rate of success. And rightly so [21], since this demand may improve IVF effectiveness, and therefore, it may reduce the costs of IVF and women's suffering [22]. It is unlikely however, that improvements in IVF could happen without research on embryo natural implantation in the uterus. Since this particular embryo research must mimic the process in the womb, there can be little doubt that this is research related with ectogenesis. This line of research has been followed at Cornell University; Prof. Hung-Ching Liu and her team have *grown a human uterus* by taking endometrial cells over a scaffolding in the shape of a uterus [23]. To test the womb, embryos left over from IVF programs were introduced, and they actually began to settle properly. The experiment was halted after six days.

There can be little doubt that this kind of experiment is research related to ectogenesis; yet the increasing demand for transparency on IVF success rates, and a competitive market, makes it improbable that this line of research would end. Moreover, coherently, either we stop demanding improvements on IVF rates of implant or we stop forbidding embryo implanting research *in vitro*.

Developments in Embryology and the Huge Gap

Neonatalogists at this point would rightly observe that with today's knowledge the gap between week 22 and the second gestational week, is insuperable. The lungs of very premature babies collapse, their blood vessels bleed easily causing damage, particularly in the brain; and infections and death are common. Yet, as they also know barriers which seem insuperable today are only temporary obstacles. Embryo development since the second half of the last century has been at the spot of arduous research. One of the most intriguing questions in biology is how a cell in a pre-embryo, 'knows' what to become, when and where has been in place for some time. Most of these questions remain unanswered; yet this remains a puzzle which embryologists are determined to resolve. For instance, Duane Alexander, Director of the National Institute of Child Health and Human Development in the 'Foreword for the Strategic Plan 2000' entitled 'From Cells to Selves' explains that the task of the Institute for the next years is to *fully understand human embryology*. In his words:

> To us falls the task not of curing a single disease or group of diseases but of solving the fundamental question in biology and all that derives from it: how does a single fertilized egg develop into a fully functional adult human being, and how do a multitude of genetic and environmental factors influence that process for good or for ill. Thus the title of this plan: 'From Cells to Selves' [24].

Amongst other areas for 'Immediate Strategic Review' in this plan of research for the new millennium appears 'Developmental Biology' which 'should understand Normal and Abnormal Development': This topic includes 'the basic biological science necessary to understand early development *in utero*, and through the time when many organ systems form' [24]. While research on human embryos after 14 days has been banned, in most countries research on animal embryos has not. Moreover, it is unlikely that the last advancements in the development of artificial wombs at both sides of gestation will remain unnoticed to Duane Alexander and his team.

The Human Genome Project and Stem Cell Research

Research on animal embryos has already shown that hormones and/or their proportion in solution may develop one organ rather than the other. Frog embryos for instance, may generate a second head when a hormone is more concentrated in the solution; alternatively, a different set of hormones may regenerate other parts of the body. In stem cell research, this knowledge has permitted cell differentiation into different types of tissue; and it has introduced the idea of developing organs for transplant derived from stem cells. In turn, findings that may arise from stem cells research may promote further advance on the knowledge leading to embryo development [25–28]. Moreover, the Human Genome Project after mapping the human genome now has started the process of decoding what a gene does, when and what for. As the Strategic Plan 2000 'From Cells to Selves' explains:

> Once the human genome is mapped, the advances will be expansive. Even within a couple of years, the mapping will provide scientists with the 'bricks and mortar' they need to understand the intricate programs for turning genes on and off – just at the right moment, in the right sequence, and triggering 'back-up circuitry' . . . [29].

These are exactly the questions relevant for the understanding of embryo development – and for ectogenesis. For some medical circles, the artificial womb is just a matter of time. Rosalie Ber, MD emeritus professor and ex-Director of Education at the Technion Medical School in Israel, for instance, writes:

> In view of the de-personalization of the gestational surrogate mother who functions as a 'womb for rent' . . . why not permit using the wombs of women in persistent vegetative state [PVS], female bodies kept viable by artificial means, *until technological perfection of an artificial womb is achieved*? [30].

Ber, appears to be *certain* that the technological perfection of an artificial womb will be achieved, eventually. Of course, she might be wrong. Nevertheless, we can hardly object to Ber's premise that 'today the technological perfection of an artificial womb is still in its elementary stages' [31].

Developments in Computer Science

Computer science has now been fully adopted by molecular biology research in order to understand protein structure and function; therefore full ectogenesis in mammals could become a reality sooner than we may think. In 2001, during the congress of the American Society of Reproductive Medicine, in Orlando, Florida, October 22–24, 2001, professor Hung-Ching Liu observed that 'her final goal is having a child in the laboratory'. In her words:

> . . . I want to see whether I can develop an actual external device . . . and then with a computer system simulate the feed in medium, feed out medium . . . and have a chip controlling the hormonal level . . . I believe if this can be achieved, we could possibly have an artificial uterus so then you could grow a baby to term [32].

Neonatalogists may address Liu's 'vision' as the views of a deranged scientist, a lonely 'cowboy'. But Liu seems to have a good reputation in her field: she is Director of the Reproductive Endocrine Laboratory at the Center for Reproductive Medicine and Infertility at Cornel University and one of the world's leading researchers in reproductive endocrinology. She has published more than 70 peer-reviewed articles in medical publications; she has delivered more than a hundred presentations on research topics at medical meetings around the world [33]. But since the conference where she reportedly made the above statements she has refused to answer to questions on this matter (Joyce Raskin, personal communication).

Facing Ectogenesis

Supposing we figure out embryo development ex-uterus in an artificial womb, safely, would it be wrong for humans to produce a child by using this device? Ectogenesis, if proved one day to be safe, might be a solution to avoid the need of a surrogate; for women who want a biological baby but lack a womb (for any reason); and for homosexual male couples. It might also, perhaps, appeal to women, who may need IVF to reproduce; women may save time, pain, bad moods and endless frustration (when embryos fail to implant for instance); generally, it may prevent damage to women's health; for although a pregnancy is not considered an illness [34] it is not a healthy job either; [35] and IVF is riskier. Even more controversial, ectogenesis may finally achieve women's equality; because a woman who wants a child might not be forced necessarily to slow down her career because of a pregnancy. And although, generally, gender inequality results mainly because of child care, and not necessarily because of the pregnancy, its roots may be found in the fact that women (unlike men) are those who bear the child and must slow down. As the last chapter of this book suggests, for some women, the option to *really* choose to bear the child (or not) in order to become a parent, could be of value.

New Conondrums

Yet the effect ectogenesis may have on current legislature (when fully developed and safe) is huge; first, as Ian Brassington notes in the next chapter, it may erode present abortion rights, which are rooted in woman's right over her body, till viability [36, 37]. Viability, due to enhanced neonatal procedures is already expanding, thus limiting the time for a woman to get an abortion. Recently, for example, David Steel, the former liberal leader who introduced Britain's modern abortion laws, has called for a dramatic reduction in the legal limit for most terminations from 24 to 12 weeks. His Abortion Act of 1967, which was regarded as one of the most significant social advances in the post-war period, legalized abortions until 28 weeks of pregnancy. In 1967 limit was cut to 24 weeks; in 1990 amid concerns that a 28-week-old fetus could survive outside the womb. Now Lord Steel reportedly feels it is time to go further 'given the advance of technology' [38]. If ectogenesis were to become a new way to have babies, viability would be possible at all times from the beginning of a pregnancy; if this were the case by present legislation in some countries (which permit abortion), women may not be allowed to abort at all.

Another trouble is that nature reacts against gross genetic errors with natural miscarriages [39] people cannot interfere since this happen 'naturally'. Yet, pulling the plug because of a possible human error in the proceedings of ectogenesis, will be considered as 'euthanasia'; which is today outlawed in most countries.

Ectogenesis will be costly; high-tech and round-the-clock care might be a very expensive commodity that only rich women (and men) could afford. Yet, most countries which permit it, surrogacy is also expensive [40]. Surrogacy agreements continue in most countries; and the ethical issues raised by this new-old way of reproduction [41] are other than the fact that only rich people can afford it. In Canada, surrogacy has been prohibited because using women as 'a womb for rent' may allow abuse of poor women. With ectogenesis however, the 'need' for a woman willing to 'rent' the womb disappears; with it also disappears the 'abusing' factor. It is unlikely however, that ectogenesis might become a procedure covered by any health insurance as it has happened with IVF, unless grounded proof appears that avoiding reproductive activity also improves women's health. This is a study which was not worth undertaking till now, certainly because people did not had a different way to have children [42].

One of the most troubling questions about ectogenesis is psychological harm to the child. This was also a question, which was asked about IVF, but it has not yet been fully addressed [43]. So far IVF children seem to be as normal as other children are. And they remain unconcerned of the method by which they were conceived. Similarly, a child who was born prematurely and had spent months in an incubator is not considered 'different' because society view this method of keeping a fetus alive acceptable. It is not implausible, therefore, to think that if society accepts ectogenesis, the ectogenic child could be considered as any other child.

Some Lessons From IVF

Back in the early seventies, IVF was considered repulsive. Some thought this is a monstrosity. If IVF is allowed, all the stabilizing threads would unravel. It was believed that IVF could threaten 'the very fabric of civilization':

> Marriage, fidelity, the essence of family; the sense of who we are and where we're headed; what it means to be human, connected, normal, acceptable; that it will threaten our ideas about love, sex, and nurturance [44].

The same words certainly apply now to ectogenesis. Yet, as pointed out by Ariel Revel in the first chapter of this volume, IVF has become almost routine. Medical insurance plans cover it, it has been used by infertile couples around the world; and by 2004, about a million test-tube babies were born [45]. The apocalyptic predictions about IVF did not realize. Yet at the time, IVF caused many to consider the possibilities of future ill-use, and it was far from acceptable. It looked as though it might never become generally tolerated, much less embraced. Nevertheless, soon after the first successful 'tube baby' birth, Louise Brown, in 1978, a poll showed that 60% of Americans thought IVF should be available for anyone who needed and wanted it.

Will this be also the case with ectogenesis? This is difficult to predict; however, as the last chapter of this volume suggests, people may accept the use of artificial wombs to solve reproductive trouble. Shortly after the birth of Louise Brown, a committee of the National Academy of Sciences in their Report Assessing Biomedical Technologies observed that the initial reaction to a given use of a technology might be very different from later reactions, should that technology become familiar over years of general use. Attitudes, the report said, have a habit of changing; people may think and feel differently in the future about marriage, procreation, or kinship and the biological family. It is, therefore, risky to predict how people might react to some of the future technological prospects [46]. As Marantz Henig observes, people have felt revulsion towards most emerging technologies, especially those that mimic functions we take to be central to our definitions of life and death, and that make us unique and human. These technologies at first often seem gruesome or barbaric in prospect, filled with technical impossibilities or ethical conundrums. Henig reminds us that blood transfusion, organ transplantation, mechanical respirators, and artificial insemination, all were greeted with suspicion; but as soon as these procedures were performed a few times, successfully, the objections faded away.

Final Remarks

The history of science, tell us that impenetrable barriers are only temporary. It is only a matter of time (and research) until someone – intentionally or by chance – finds a way to overcome an obstacle. Although in the case of the artificial womb there seem to be yet too many barriers, it would be naïve to suppose that things may develop differently than in past scientific breakthroughs. I have pointed out in this chapter that it is time to acknowledge fully the consequences of new developments in different areas of scientific research, which are now converging into the development of an artificial womb. Therefore, a serious discussion of this topic should start now, while we have still enough time to decide what we may want and why. A starting point for such discussions can be found in the next three chapters of this book.

Acknowledgments Research for this paper was carried out during a Visiting Fellowship at the Hastings Center in summer 2004. A version of this chapter was published in *Philosophy of Medicine and Health Care* 2006 (9) 359–365.

References

1. Singer P. and Wells D. 1984. *The Reproduction Revolution. New ways to Making Babies.* Oxford: Oxford University Press.
2. Harris J. 1998.*Clones, Genes and Immortality*. Oxford: Oxford University Press.
3. Strong C. 2002. Overview: a Framework for Reproductive Ethics. In *Ethical Issues in Maternal-fetal Medicine.* Ed. Dona L. Dickenson. Cambridge: Cambridge University Press, pp. 17–36.
4. For example, Julian Savulescu at the International Congress on the Ethics of IVF in London during September 2004 asked the scientific panel to refer to ectogenesis. The panellists just ignored the question.
5. Ber R. 2000. Ethical Issues in gestational Surrogacy. *Theoretical Medicine and Bioethics* 21: 153–169.
6. Singer P. and Wells D. 1984. *The Reproduction Revolution. New ways to Making Babies.* Oxford: Oxford University Press.
7. As Peter Singer and Dianne Wells remarkably predicted two decades ago. While Singer's thought experiment on ectogenesis was developed to support an argument about abortion and was *not* related directly to the technological progress in that direction, they nevertheless, based their idea on a linear projection of further developments in neonatal care.
8. Eben D. 2004. Numbers on Premature Babies. Saturday supplement *Maariv newspaper* (in Hebrew) 16.7.04 page 2.

9. In Denmark, for example, consensus has been reached against treating these babies. However, in other countries this is not the case. Saviour acts to keep alive premature babies are highly influenced by pro-life campaigners. In addition, some neonatalogists observe that they cannot know *a priori* which one of these babies may survive and even get to college (personal communication).
10. Jennings B. 1988. Beyond the Right of the Newborn. *Raritan* 7: 79–93.
11. Sills J. 2003. Understanding Catastrophic Health Care Exposures. Neonatal Intensive Care – How did we Get Here and Where are we Going? *American Re-HealthCare Symposium 2003*. http://www.amre.com/hc2003/summaries/sills.htm. accessed 23.03 2004.
12. http://www.w-cpc.org/news/reuter7-97.html. accessed 13.7.04.
13. The first special unit for premature infants was opened in 1922; but in 1988 there were already 600 hospitals in the United States providing 7,500 NICU beds [10]. Paradoxically, the remarkable success of ARTs is the main contributor to NICUs: in Israel, for instance, almost half of the VLBW infants admitted to NICUs are the product of ART, and more than a third of VLBW live-born infants are multiple births. VLBW (birth weight less than 1,500 gram) rates increased from 1.1 percent to 1.3 percent of the total live births from 1995 to 1998. Between 1993 and 1998, the total number of VLBW infants born each year increased by 40 percent Zmora Ehud. 2001 Ethics and Neonatalogy in Israel. *The Journal of Clinical Ethics* 12: 304–307.
14. Sills J. 2003. Understanding Catastrophic Health Care Exposures. Neonatal Intensive Care – How did We Get Here and Where Are We Going? *American Re-HealthCare Symposium 2003*. http://www.amre.com/hc2003/summaries/sills.htm. accessed 23.03 2004
15. Ly H. 2003. Perinatal care at the threshold of viability – from principles to practice. *Ann Acad Med Singapore* 32: 362–375.
16. Lorenz JM. 2001. The outcome of extreme prematurity. *Semin Perinatol* 25: 348–359.
17. Lee SK, McMillan DD, Ohlson A, et al. 2000. Variations in practice and outcomes in the Canadian NICU network: 1996–1997. *Pediatrics* 106: 1070–1079.
18. Piecuh RE, Leonard CH, Cooper BA, Sehring SA. 1997. Outcome of extremely low birth weight infants (500–999 grams) over a 12-year period. *Pediatrics* 100: 633–639.
19. *Proceed with Care* Final Report of the royal commission on new reproductive technologies 1993 minister of government services, Canada. Ottawa: Canada Communications Group. *Recommendation* 184: 637.
20. *Proceed with Care* Final Report of the royal commission on new reproductive technologies 1993 minister of government services, Canada. Ottawa: Canada Communications Group. *Recommendations* 110–112: 542.
21. Improving rates of IVF in the US has also become a competitive target in a free market.
22. The rate of implantation is low in nature as well. Yet the natural process is cheaper and painless.
23. The scaffolding dissolved as the cells grew into uterine tissue. The womb was then supplied with proper nutrients and hormones.
24. Alexander D. 2000. Foreword From Cells to Selves Strategic Plan 2000 National Institute of Child Health and Human Development. http://www.nichd.nih.gov/strategicplan/cells
25. Wilmut I and Dominsky T 2000. Editorial. Government Encouragement for Therapeutic Cloning. *Cloning* 2: 53–54.
26. Ahn JI, Lee KH, Shin DM, et al. 2004. Comprehensive transcriptome analysis of differentiation of embryonic stem cells into midbrain and hindbrain neurons. *Developmental Biology* 265: 491–501.
27. Brandenberger R, Wei H, Zhang S, Lei S, Murage J, et al. 2004. Transcriptome characterization elucidates signaling networks that control human ES cell growth and differentiation. *Nature Biotechnology* 22: 707–716.
28. See for example some of the implicit targets in Wilmut and Dominsky [25]. Ongoing research in stem cells, which searches developmental expression of genes, can be found in Ahn et al. [26] and Brandenberger et al. [27]. This list is not exhaustive.
29. Strategic Plan. 2000. From Cells to Selves.

30. Ber R 2000. Ethical Issues in gestational Surrogacy. *Theoretical Medicine and Bioethics* 21: 153–169.
31. Ber R 2000. Ethical Issues in gestational Surrogacy. *Theoretical Medicine and Bioethics* 21: 154.
32. OBGYN.net Conference Coverage http://www.obgyn.net/. accessed 13.7.04
33. http://www.ivf.org/liu.html accessed 13.7.04.
34. Cook R, Dickens BM, and Fathalla MF. 2003. *Reproductive Health and Human Rights*. New York: Oxford University Press.
35. Reproductive hazards have traditionally been viewed as women's fate and taken for granted. In this view, Cook, Dickens and Fathalla explain candidly that "maternity is not a disease" but, "an essential function that women fulfil for the survival of our species" [35, p. 29] The problem with this definition of maternity is that although this is not a disease, this is not exactly a very 'healthy job' either.
36. http://www.canadianlawsite.com/abortion-laws.htm
37. There are differences from country to country. For example, in England two doctors have to agree that the mother's mental or physical state could be damaged by continuing with the pregnancy. http://www.nhsdirect.nhs.uk/; In Europe some countries are more liberal than others. However, in Cyprus, Israel, Poland, Portugal, Spain and Switzerland, abortion laws are still very restrictive, allowing abortion only in cases of rape or foetal impairment, or to protect a woman's physical or mental health. http://www.ippf.org/regions/europe/choices/v28n2/legislation.htm.
38. Watt N 2004. Steel calls for abortion limit to be cut. *The Guardian* July 5 http://society.guardian.co.uk/ accessed online 5.7.04.
39. Marchetti F, Bishop JB, Cosentino L et al. 2004. Paternally Transmitted Chromosomal Aberrations in Mouse Zygotes Determine Their Embryonic Fate. *Biology of Reproduction* 70: 616–624.
40. A British homosexual male couple that hired a surrogate mother in the US to deliver them twin babies, for example, were able to do so because they had the money to pay for it.
41. Sarah used Agar to have a baby from Abraham; today in countries where the number of children per woman remains high, women may give one of their own children to a barren sister (personal communication).
42. Moreover, women had babies, even when it was a highly deadly adventure. That is before hygiene, antibiotics and cesareans became available. Most probably, and in most cases, it may not have been women's choice to become pregnant in the first place. However, it is quite improbable that women's societal conditioning over the eons towards a gendered biological predetermination would suddenly end in a full stop. Plausibly, most women will choose 'the old way' of making babies, at least once.
43. In the recommendations section of the last report of the President's Council on Bioethics, the council recommends to do a federally funded longitudinal study of the impact of ARTs on the health and development of children born with their aid. A Report of the President's Council on Bioethics. 2004 Reproduction and Responsibility. *The Regulation of New Biotechnologies*. Washington, D.C. www.bioethics.gov. p. 208.
44. Marantz Henig R 2004. *Pandora's Baby*. New York: Houghton Mifflin.
45. *2004 The Associated Press*. http://msnbc.msn.com/id/3088015/ accessed 29.6.04.
46. Marantz Henig R 2004. *Pandora's Baby*. New York: Houghton Mifflin.

Chapter 14
Human Before Sex? Ectogenesis as a Way to Equality

Tuija Takala

Introduction

Within Gender studies in particular and in many other academic disciplines in general, the simplistic dichotomy between men and women has been long refuted. In the real world, however, boys are still brought up to be men and girls to be women with the corresponding role expectations. Even in the more enlightened circles, women's role and place is, if not determined, definitely informed by the fact that they could be, or could become, mothers. A news item in the biggest broad sheet newspaper in Finland celebrating the International Women's day on 8 March 2008 included interviews with teenage girls. One of the interviewees told the journalist that she liked being a girl because that allowed her to dress up and that she waited to be a mother because only that will transform her from a girl to a woman [1]. I found that very alarming. If teenagers of our time living in European capital cities think that they need to become mothers to become women, not much has changed in the last hundred years. Women do not need to blame only men for the hardships they face, as they (I refuse to write "we") are confining themselves to specified roles.

What I will propose in this paper is nothing new. The same suggestion was made by Shulamith Firestone in her book *The Dialectic of Sex: A Case for Feminist Revolution* (1970) [2]. In her book Firestone argued that women are oppressed by the patriarchy through their biology; the physiological, social, economical and psychological disadvantages caused by pregnancy, childbirth and caring for the children. While I do not share all of the Marxist undertones of Firestone's theory, I am in agreement with her that the reason for the continuing lack of equality between the sexes boils down to the fact that women (and only women) are expected to bear and rear children. At the time *The Dialectic of Sex* was published, the possibility of ectogenesis was still somewhere far in the future, but this is no longer the case. Contrary to what many feminists have argued my claim is that women, collectively and individually, should welcome the possibility of nurturing the embryo and foetus outside the womb for the length of the entire gestation period. In this paper I will

T. Takala (✉)
Department of social and moral philosophy, University of Helsinki, Finland
e-mail: tuija.takala@helsinki.fi

F. Simonstein (ed.), *Reprogen-ethics and the Future of Gender*, International Library of Ethics, Law, and the New Medicine 43, DOI 10.1007/978-90-481-2475-6_15,

show how having the option of ectogenesis is in the interests of all females and that if and when ectogenesis becomes a safe option it will finally make true equality between humans possible.

Ectogenesis

In the feminist and bioethical literature the term ectogenesis is used to refer to a number of ways in which the species typical pregnancy (gestation in the womb) is substituted by alternative means. For some, the term ectogenesis refers solely to the use of artificial or mechanical wombs [3–5], whilst many others take it to include also the possibilities of creating womb-like conditions elsewhere inside the body; a body that could be either male or female [6]. The term "ectogenesis" in itself combines the words "ecto" which roughly means "outside" and "genesis" which means "origin" or "start", and as such, of course, allows both definitions. Some authors, while recognising that technically ectogenesis can be taken to mean simply outside the womb, choose to concentrate their analysis on artificial wombs [7] or on gestation taking place outside a woman's body [8]. In what follows, I will use the word ectogenesis to refer to artificial wombs outside a human body, unless otherwise stated. I will discuss ectogenesis primarily as an alternative to the whole gestation process, and will not talk about partial ectogenesis more than simply stating here that, obviously, if total ectogenesis can be perfected, it will provide instruments for saving premature babies too.

Although the technical possibilities of ectogenesis are not quite with us yet, already the previous century saw three significant ectogenesis debates. As early as in the 1920s the topic was debated in the UK and in that debate many of the modern viewpoints were put forward. Some argued that ectogenesis would enable humanity to control reproduction and to improve the species, and that by allowing people to separate sex from reproduction it would increase people's possibilities to engage in sexual pleasures. Others saw the idea of ectogenesis as a ploy of body-hating feminists whilst some would have welcome it as a way of liberating women. Unsurprisingly, there were also people who were concerned about what this would do to traditional family structures and many worried that "artificially made" humans would somehow become "artificial" [5].

In the 1970s and 1980s the debate peaked within the feminist circles. Much of this debate was sparked by Firestone's book *The Dialectic of Sex* and subsequent commentators either agreed with Firestone that ectogenesis would liberate women or, alternatively, saw ectogenesis as a further means to control women and demean them. Those more hostile towards the possibility of ectogenesis saw it as a development that could ultimately make women obsolete [5].

In 1985 a chapter on ectogenesis by Peter Singer and Deane Walls in their book *Making Babies: The New Science and Ethics of Conception* added bioethical elements to the debate [9]. By this time ectogenesis was no longer just a sci-fi scenario in the sense that smaller and smaller premature babies could be kept alive in incubators outside the womb with the help of various technological devises. As Singer and Walls noted; "[t]he period in which it is necessary for the human foetus to be in its mother's womb is shrinking from both sides" [9]. In their paper they put forward five arguments which might be made in favour of ectogenesis. These have been called the "better surrogacy argument," the "abortion reconciliation argument," the "sexual equality argument," the "better parenting argument" and the "partial ectogenesis for transplant argument" [10]. These arguments by Singer and Wells were constructed to show how people with different normative presuppositions should, for different reasons, come to accept ectogenesis. The arguments put forward in their paper should not be taken as a unified case for ectogenesis, but they should rather be seen as a collection of reasons (some of which should convince some people and others other people holding different normative presuppositions) that perhaps there is something ethically desirable in ectogenesis. In trying to justify ectogenesis from a number of directions, they also came to explicate arguments that have since then been used against ectogenesis. For instance, by making the argument that ectogenesis could be used to save the lives of aborted foetuses, they opened the door for claims that women's reproductive rights could decrease if there was a way of keeping the aborted foetuses alive [5, 10, 11]. Further, by introducing partial ectogenesis as a way of producing spare parts for existing people, they were sure to create a number of "yuck-reactions" some of which could even stand after calm and dispassionate analysis. I will return to some of these arguments later in this chapter, but at this point, it is interesting to note that in the discussion that immediately followed, Singer and Wells's argument concerning equality was not debated [10, 12]. And equality, I claim, is the most important reason for arguing for ectogenesis.

Depending on whose scientific opinion you listen to, artificial wombs capable of sustaining a foetus's life throughout the nine months' gestation period could become a reality in a couple of years' or in a couple of decades' time [6, 13]. I will not look at the science in any detail nor will I discuss the problems relating to research that needs to precede the availability of artificial wombs. My point is to argue for the general desirability of ectogenesis. As my main thesis is that ectogenesis would open possibilities for equality never feasible before, I will look at the feminist discussion, which supposedly should be interested in such matters, in some detail.

Obviously, there is no one feminist stand on the issue of ectogenesis. Liberal, radical, eco, Marxist, post-modern and other schools of though within feminism all start from slightly different premises and reach dissimilar conclusions and there is disagreement within the groups also. In what follows, I will have a closer look at themes that are frequently discussed in the feminist debates whatever the more specific framework.

Ectogenesis is in Everyone's Interests

If a teenage girl thinks that she needs to become a mother in order to become a woman [1], there is something seriously wrong with the role models we are giving to the younger generations. To start with, not all women are capable of reproducing.

They could be born infertile, they can become infertile later in life for a variety of reasons, they might not find a partner to reproduce with, or they might not have access to suitable fertility treatments. Also, an increasing number of people have ideological reasons for deciding not to reproduce [14–17]. If ectogenesis were available and in use, none of these people would have to feel incomplete – the assumed, almost necessary, connection between female genitalia and motherhood would have been broken.

Breaking the link between being a female and being a mother is, I would argue, in everyone's interests. Those females who cannot reproduce will no longer have to feel incomplete. Those who wish to take on motherhood can be praised and thanked for their contribution, as motherhood is no longer what was expected of them anyway. Women would have a choice as to whether they would like to take on all the risks associated with natural pregnancy and put themselves through all the discomforts and pains related to even to pregnancies where everything goes well, or whether they would like to reproduce without subjecting their bodies to all this – if reproducing is what they want to do [18].

If reproduction no longer were only women's business it would help a number of people not currently quite fitting into the dichotomies of males and females, men and women. Biologically speaking as many as 2% of children are born with ambiguous genitalia [19], and many more have chosen an identity not quite fitting the simplistic division between the genders; a division which still flourishes because of the insistence on the biological determinants. For people currently in the margins, life in the world of ectogenesis would be considerably easier.

Females have had the privilege (for those who see it as such) to bear children, because it has been physically possible only to them and because the womb has provided the best possible place for the foetus to develop. When neither of these is no longer true, why should women maintain the privilege? If equality is what we are after, there is no reason to allow one group of people a privilege simply because they can. There are many evolutionary instincts that people have given up to make room for social evolution, so why should the need of (some) females to bear children be any different?

I suppose the strong opposition to ectogenesis coming from many feminist stems from the fact that it might increase the rights of males when it comes to reproduction. But for anyone who is truly for equality this should not be an unwelcome development. The reason that women have traditionally had a bigger control over reproduction has been that the gestation has taken place within the female body and that most reproductive decisions so far have included interventions to that body. However, already in cases of fertilised embryos resulting from *In Vitro Fertilisation* (IVF) questions have been asked whether women should legitimately always have the last say [20]. In my utopia we would all be human beings first and only through choice would we remain childless, become social or biological parents, or, in the case of some females, biological mothers. Perhaps as our knowledge of the human gestation increases, it might become possible to rear children within a human body, but without the need of a womb. This would open the option of biological "motherhood" for all humans. However, even this scenario would give females more choice, as they (and only they) would still have the option of the species typical biological motherhood that, if all goes well, requires no artificial interventions.

Some feminists would still be worried that taking away females' monopoly on reproduction would marginalise women even further, perhaps even to the point of making them obsolete [5]. The idea that women must maintain monopoly over producing offspring is sometimes framed as the "elimination of women argument" [21]. The feminist literature is divided in whether they see the female biology as inherently debilitating and something from which women should be glad to escape or whether the body is viewed as valuable and potentially empowering [8]. Many intermediate positions exist too, according to which female biology is potentially debilitating and as such women should have control over it (for instance, birth control, right to abortion and the right to choose the method of delivery), but that female biology is also a source of power (because it has the monopoly over producing offspring) and as such it should not be forfeited.

The worry seems to be that if women were no longer necessary for reproduction, the male dominated world would no longer need the females, and thus by advocating ectogenesis, women are actually advocating their own extinction. Still, many feminists would acknowledge that women's oppression results at least partly from their biology; of which not an insignificant part is the bearing and rearing of children that limit women's ability to contribute to society's other functions. First of all, I would claim, like many others, that those with female genitalia do have a lot more to offer than their function as "baby machines" [9, 21]. I would, however, like to go even further and claim that since the oppression of women was made possible by females' role as mothers, because these roles prevented women from having an equal part in communal life, the possibility of ectogenesis would not only free women from pregnancy, but that as it would take away the biggest difference between the sexes, it would also take away the grounds for oppression and eventually, lead to true equality. As there would no longer be reasons for the differential treatment of men and women because of their sex-specific reproductive roles, the structures of oppression would, with time, come crumbling down.

What About the Children

Ectogenesis is not only about the rights of existing people, because its function is to produce new people and most would think that their interests should also somehow be accounted for. When it comes to ectogenesis some feminist seem to have a somewhat schizophrenic attitude to the best interests of the child. On the one hand, ectogenesis is criticised because it would break the natural bond between a mother and her child, and because children gestated in artificial conditions would be deprived of all the good things that only a natural womb can provide. What these exactly include and whether any of this can be proven is another question, but basically the point is that since the womb provides the best conditions for a foetus to develop, ectogenesis is bad. On the other hand, however, when the same author discusses a possible future where ectogenesis is perfected to the degree that it could actually provide better conditions for a foetus to develop than a womb would, that would be bad, because it might interfere with a woman's right to reproduce [8]. I do not see how best conditions for a child to develop can both be seen as a reason for being against something and as a reason that needs not to be taken into account. Further, one might think that the good or the virtuous mother celebrated by some feminists [22] would be primarily interested in what is best for her child. Suffice it to say that obviously, before ectogenesis can provide a safe and at least as good a start for a human life as a natural pregnancy would, it would be wrong to use it. Then again, other feminists seem to make pregnancies primarily about the mother, the child being just a by-product of the essentially female experience [7]. This, to me, is a selfish position beyond all justification.

In *The Dialectic of Sex* Firestone controversially claimed that ectogenesis would liberate children from over-possessive mothers and as such lead to healthier mother and child relationships, because without the nine-month pregnancy, discomfort and painful delivery women could no longer claim that the child somehow "belonged" to her [2, 9]. Others have actually welcome the opportunity to claim authority over their children by being able to claim that the children owe something to their mother because of the hardships she went through in bearing and rearing them [5]. Since the children in question did not ask to be born, but were, in most cases, brought to this world because their mother wanted to have a child, I find this playing the guilt card unjustified. And even if the mother was given no choice, it is not the fault of the children that their coming to the world caused discomfort and pain to their mother. More generally, though, the worry seems to be that if ectogenesis becomes more common, the basic human relationship between a mother and her child disappears and that could have an adverse effect on society as a whole [8]. I wonder, though. The mother-child relationship is often characterised by possessiveness, partiality and unreasonable expectations, but these are not the sort of features that societies particularly need. Life in communities bigger than a family unit or a small village requires impartiality and justice rather than the abovementioned attributes. Not only am I inclined to agree with Firestone that ectogenesis could contribute towards healthier mother-child relationships, but I would go even further to argue that society as a whole would benefit.

Adverse Consequences of Ectogenesis

I have already briefly touched upon the fear that as the methods of ectogenesis develop, it could eventually become the only option of reproduction available. It is, after all, conceivable that at some point the well-being of the future child could best be guaranteed in the controlled environment of the artificial womb and that this would lead to a situation where natural pregnancies were first discouraged and later perhaps even condemned. Such a scenario might indeed limit women's choice, but how bad a development this would be is another question. Even if experiencing pregnancy is intrinsically valuable to some women, it is difficult to make a convincing argument stating that that value is so great that it would justify compromising the well-being of the resulting child. If ectogenesis was indeed shown to be in the best interests of the children and we truly believed in equality, this reduction in choice (by taking away a potentially harmful privilege) should not be seen as an adverse effect. Some, however, have voiced qualms that it would take even less for the society to coerce women to choose ectogenesis than the best interest argument; that the availability of ectogenesis would change our attitudes towards natural pregnancy, which would become viewed as primitive and disgusting [8]. All I can say is that coercing for such reasons would not be justified and given that I believe that the availability of ectogenesis will change the existing power balance towards a more equal model, I would claim that such a scenario is unlikely to become a reality.

A related worry is that through ectogenesis we will learn more about the foetal development, which could then be used to justify greater control on normal pregnancies [21]. While it is undoubtedly true that research into ectogenesis and the practice itself will make us more aware of what happens during foetal development and that this could cause pressure to increase control on women's behaviour (nutrition, environments etc.) during pregnancy, similar information will be gathered without ectogenesis too. For instance, research into ways of saving premature babies as well as embryo research contribute to our increasing understanding of foetal development. Similar arguments as presented in the above could be used here, but in terms of discussing the pros and cons of ectogenesis, this line of argument is ineffective and I will not pursue it any further.

There is, however, also the issue of what happens to abortion rights when ectogenesis becomes a reality. The problem for the person making the abortion decision is that once ectogenesis is perfected (and methods of abortion adjusted for the safe removal of the foetus), the decision to abort could be viewed as a choice between giving my newborn child for adoption and letting it die. Also, when there is a possibility that the aborted foetus could survive, the woman is likely to feel a maternal sense of responsibility for the foetus [22]. What this means is that the decision to abort (without the intention of making use of ectogenesis) becomes more difficult than decisions to abort were before the technology existed. Also, in terms of legal rights to abort, there will be problems in justifying allowing foetuses that could survive die. And assuming that ectogenesis will be relatively expensive and that there will not be enough adoptive parents to take all the foetuses that have been raised in artificial wombs for adoption, it is possible that abortion could become far more strictly regulated than it currently is [23]. However, similarly to the above cases, the advances in neonatal care will eventually lead to similar problems. With or without total ectogenesis, the moral status of the foetus and related ethical and regulatory issues will require much discussion within the next few years.

Many of the adverse effects mentioned in relation to ectogenesis are problems we will face without intentionally aiming for ectogenesis. All of them raise difficult moral and regulatory problems, which, need to be addressed; but as they are not exclusive to ectogenesis, I will not attempt to answer them here in any more detail.

Concluding Thoughts

Since the beginning of time females have got pregnant, borne children, given birth, nurtured and reared the children. While methods of birth control and various reproductive technologies have increased women's abilities to plan their pregnancies and the increased equality has allowed women a fuller part in the life outside the family, pregnancy has remained the woman's job. It is only now, with the possibility of ectogenesis, that the allegedly profound distinction between those who could, in principle, be or become mothers and those who cannot, will be challenged. Women should embrace the possibility of ectogenesis as a road to true equality. And not only women, but also all those who do not fit in the simplistic stereotypes of men and women, as these categories will gradually lose their importance. For men, ectogenesis will increase their reproductive rights also with a variety of new roles that become possible for them. With the development of ectogenesis we could finally be freed from the biologically determined roles that have for long dragged behind the social evolution. In most other areas of life humans have risen beyond their biological instincts, and now it could happen with reproduction also. This would mean profound changes in societies, but most importantly, it would make us all first and foremost human – not bearers of roles that we might not see as our own.

References

1. *Helsingin Sanomat*. 8 March 2008. Page C12.
2. Firestone, Shulamith. 1970. *The Dialectic Sex: A Case for Feminist Revolution*. New York: Farrar, Straus and Giroux.
3. Shanner, Laura and Nisker, Jeffrey. 2001. Bioethics for clinicians: 26. Assisted reproductive technologies. *Canadian Medical Association Journal* 164: 1589–1594.
4. Pence, Gregory. 2006. What is so good about natural motherhood? (In praise of unnatural gestation) In *Ectogenesis: Artificial Womb Technology and the Future of Human Reproduction*, eds. Scott Gelfand and John R. Shook, pp. 77–88. Amsterdam and New York: Rodopi.
5. Tong, Rosemary. 2006. Out of body gestation: In whose best interests? In *Ectogenesis: Artificial Womb Technology and the Future of Human Reproduction*, eds. Scott Gelfand and John R. Shook, pp. 59–76. Amsterdam and New York: Rodopi.
6. Simonstein, Frida. 2006. Artificial reproduction technologies (RTs) – all the way to the artificial womb? *Medicine, Health Care and Philosophy* 9: 359–365.
7. Aristarkhova, Irina. 2005. Ectogenesis and mother as machine. *Body & Society* 11: 43–59.
8. Sander-Staudt, Maureen. 2006. Of machine born? A feminist assessment of ectogenesis and artificial wombs. In *Ectogenesis: Artificial Womb Technology and the Future of Human Reproduction*, eds. Scott Gelfand and John R. Shook, pp. 109–128. Amsterdam and New York: Rodopi.
9. Singer, Peter and Deanne Walls. 2006. (orig. 1985). Ectogenesis. In *Ectogenesis: Artificial Womb Technology and the Future of Human Reproduction*, eds. Scott Gelfand and John R. Shook, pp. 9–25. Amsterdam and New York: Rodopi.
10. James, David N. 1987. Ectogenesis: A reply to Singer and Wells. *Bioethics* 1: 80–99.
11. Cannold, Leslie. 1995. Women, ectogenesis and ethical theory. *Journal of Applied Philosophy* 12: 55–64.
12. Wells, Deane. 1987. Ectogenesis, justice and utility: A reply to James. *Bioethics* 1: 372–379.

13. Gelfand, Scott. 2006. Introduction. In *Ectogenesis: Artificial Womb Technology and the Future of Human Reproduction*, eds. Scott Gelfand and John R. Shook, pp. 1–7. Amsterdam and New York: Rodopi.
14. Häyry, Matti. 2004. A rational cure for prereproductive stress syndrome. *Journal of Medical Ethics* 30: 377–378.
15. Benatar, David. 1997. Why it is better never to come into existence. *American Philosophical Quarterly* 34: 345–355.
16. *Childless by Choice*. http://northvalley.net/cbc/
17. *The Voluntary Human Extinction Movement*. http://www.vhemt.org/
18. Smajdor, Anna. 2007. The moral imperative for ectogenesis. *Cambridge Quarterly of Healthcare Ethics* 16: 336–345.
19. Blackless, Melanie. et al. 2000. How sexually dimorphic are we? Review and synthesis. *American Journal of Human Biology* 12: 151–166.
20. Bennett, Rebecca. 2008. Is reproduction women's business? How should we regulate regarding stored embryos, posthumous pregnancy, ectogenesis and male pregnancy? *Studies in Ethics, Law and Technology* 2(3): article 3.
21. Murphy, Julien, S. 1989. Is pregnancy necessary: Feminist concerns about ectogenesis. *Hypatia* 4: 66–84.
22. Cannold, Leslie. 1995. Women, ectogenesis and ethical theory. *Journal of Applied Philosophy* 12: 55–64.
23. For further discussion on terminations and ectogenesis, see Iain Brassington, Chapter 15, Section "Terminations and Glass Wombs" in this book.

Chapter 15
The Glass Womb

Iain Brassington

Introduction

A Bill presented before the legislature of the State of Ohio on the 19th July 2007 proposed that the termination of a pregnancy should be permitted to go ahead only with the consent of the father of the foetus [1]. Specifically, under division 1(B) of the Bill,

1. When the foetus that is the subject of the procedure is viable, no person shall perform or induce an abortion on a pregnant woman without the written informed consent of the father of the foetus.
2. When the foetus that is the subject of the procedure is not viable, no person shall perform or induce an abortion on a pregnant woman without the written informed consent of the father of the foetus.

And in division 1(C)(2), it is stipulated that "[n]o pregnant woman seeking to abort her pregnancy shall fail to comply with division (1)(B) of this section". The Bill, if passed, would do more than formalise a system of paternal veto for terminations of pregnancy: division 1(G) states that "[i]t is not a defence to a violation of division (B)(1) or (2) or (C)(2) of this section that the woman does not know the identity of the father of the foetus". It is not just the case that the father has a right to forbid a procedure that would otherwise go ahead: the procedure could not be initiated without his say-so. A woman who cannot name the father cannot obtain his consent; therefore, she would not be able to have the procedure legally.

For those who think that abortion is morally impermissible, this Bill might be something to be welcomed as making the impermissible harder to obtain. For many who are not opposed to abortion – and especially for those who think it is a right – the Bill might seem to have missed something important. The nature of this something would be related to an agent's autonomy and sovereignty over her body: I shall return to this point in a moment. But there is one other response to which we should pay attention, too: that, even if there is nothing especially wrong with abortion, and

I. Brassington (✉)
CSEP/iSEI, School of Law, University of Manchester, Manchester, England, UK
e-mail: iain.brassington@manchester.ac.uk

F. Simonstein (ed.), *Reprogen-ethics and the Future of Gender*, International Library of Ethics, Law, and the New Medicine 43, DOI 10.1007/978-90-481-2475-6_16,

even if there is some kind of right to it, the Bill might, for all its failures, have hit on something important. And if this last response is tenable, serious questions are raised in connection with new and possible reproductive technologies.

A Rationale for the Bill

For many, the thing that has gone wrong with the Bill is that it seems to undermine agentic sovereignty over the body and moral decisions. Moreover, it does so in a way that – intentionally or not – keeps the female body tacitly but firmly under male control.[1] Thus, straightforwardly enough, the Bill is problematic. Moreover, accepting that the Bill is problematic does not mean that one necessarily thinks abortion permissible: one might think abortions morally problematic, but hold at the same time that what is more problematic (at least until a certain stage in the pregnancy[2]) is that an agent should no longer have sovereign control over his or her body. Such a view might be defended in a number of ways. The details of such defences are not important here – suffice it to say that we could reach relevantly similar conclusions by adopting either broadly Millian appeals to liberty [2] or broadly Kantian appeals to autonomy [3]. Maybe we would want to live in a world in which people did not have abortions – but this may not suffice to justify the Bill if sovereignty comes up trumps.

And yet it might also be the case that there is a couple of rationales for the Bill. One of these is straightforwardly "pro-life". A more explicitly anti-termination Bill would be legally difficult: *Roe v Wade*, which established that a woman does have the legal right to a termination of pregnancy, sets the constitutional scene in the USA, and therefore in Ohio. Strictly speaking, the Bill would preserve the mother's legal right to terminate her pregnancy. But there is no reason why one should not append to this right the claim that certain conditions must be met (her being competent to give consent for the procedure, being able to pay for it or have it paid for, and so on) in order for that right to be exercised; and all one would be doing in endorsing the Ohio Bill is adding another condition. If the father is happy for the termination to go ahead, then it may. Nevertheless, just because terminations would be more difficult to obtain, one suspects that insistence that the father of the foetus must consent to a termination of pregnancy ticks many of the boxes that pro-lifers would want ticked by maximising the number of opportunities for the prevention of an *ex hypothesi* wrongful action, and thereby minimising the chance of it actually going ahead.

On the other hand, deliberately to make it difficult to obtain that to which someone is legally entitled is, perhaps, to display a dog-in-the-manger attitude; and the mere fact that we disapprove of something morally is not sufficient to warrant its

---

[1] It may or may not be coincidental that eight of the nine sponsors of the Bill are male.

[2] That is, we might think that at a certain point – viability, say – the foetus does have a serious right to life that has to be considered. Still, we could accept that, before that point, the mother is accountable to no one but herself for what she does to her body and its occupants.

being legally forbidden – assuming, of course, that we are in the right to disapprove of that something in the first place. For the purposes of this paper, I shall assume that the foetus has no especial right to life, and that the pro-life rationale for an instrument like the Ohio Bill therefore fails.

However, although one might suspect that the Bill would be supported by a good many of those who would describe themselves as pro-life and that the motivation behind its introduction was pro-life as well,[3] an instrument such as the Ohio Bill could be supported by others outside of, and perhaps even hostile to, the pro-life camp. All that is necessary is to hold to the thought that, even if we allow that terminations of pregnancy are permissible, the mother of the foetus does not have the power to authorise them unilaterally. Holding this position might indicate that we are lending support to male hegemony over the female body; but it need not. It might reflect an adherence to a set of claims about parental partnership.

Parenthood is often described as a joint project; and if this description is accurate, we could legitimately expect that each parent would be entitled to some kind of say in the progress of that project. In other contexts, we might say that two people who set up a company both have a say in how it is governed, or that people who share the cost of a house both have a say in what colour to paint the spare bedroom. Since one is no less a parent for the foetus not having been born yet, it would seem plausible to suppose that, *ceteris paribus*, fathers ought to have a say in decisions concerning the termination of a pregnancy. Adhering to such a position would not commit us to the view that an abortion would be impermissible; it would simply reflect the view that an abortion without authorisation would be impermissible, for the same reason that it would be impermissible for one of the partners in a business to close it without the authorisation of the other. Both biological parents, we might say, have a stake of the right kind in the foetus to mean that both ought to authorise the termination of a pregnancy.[4]

Such a rationale fails – but it is not utterly ridiculous, and I am satisfied that it does touch on something important that ought not to be forgotten. Were the line to be successful, claims about agents' sovereignty would be of limited success against it, just because it draws from the initial supposition that having a child is a joint project, which implicitly means that the parents willingly surrender at least some of their sovereignty to that project. (To continue the analogy with a business: if one director sold the warehouse while the other was on holiday, he could not really complain that his (presumably erstwhile) partner lacked respect for his sovereignty when he returned and began legal action.) If something is a joint project, then we can expect that decisions be made either jointly or, if individually, only within the confines of some jointly-settled rubric.

---

[3] I have emailed the sponsors of the Bill to ask what their motivation was, but – perhaps unsurprisingly – have had no reply.

[4] It may be worth noting that it would be politically expedient to downplay the pro-life rationale for the Bill in comparison to the "joint project" rationale, just because a person may think both that terminations of pregnancy are permissible, and also that they require the authorization of both parents. Were the Bill to forbid terminations outright, it would not be able to rely on the support of the constituency that does think in this way.

Yet the "joint project" line does fail. Not only does it fail rather quickly; it also fails for a fairly obvious reason: its reliance on all other things being equal when, in real life, all things are patently not equal.[5] While there ought to be no problem accepting that both parents have an equal genetic stake in the foetus, that they may well have an equal emotional stake in it, and, assuming that the pregnancy was embarked upon deliberately, that both have an equal teleological stake in it (insofar as it is a constituent part of the developing family and their desired status as parents), it is also obviously true that the mother is physically attached to it in a way that the father is not. And this physical proximity is very important morally. If we think that abortions are permissible, there is a very good reason also to think that the mother of a foetus has an unassailable priority in deciding whether or not to continue with her pregnancy. Moreover, if we do not already think that abortions are permissible, appeals to the moral value of having control over one's body give us a pretty good reason at least to consider changing our mind. This is the kind of thought exploited by Thomson in her famous violinist example [4].

Finally, if, having revisited Thomson, we are steadfast in our belief that it is wrong to terminate a pregnancy, we would still lack a reason to attach any particular importance to the father's consent or lack of consent. For if a woman who deliberately acts to end her pregnancy has committed a wrong, that wrong is separable from the putative wrong of acting without the father's approval. After all: if we think that abortion is wrong and the kind of thing that might justly be prevented, it is not clear what merely being the father adds in such a case, nor why his acquiescence would allow us to permit the procedure after all. (Go back to the business analogy: directors ought not to raid the pension fund, and they ought to keep each other informed about their activities. But raiding the pension fund is not made any worse by not getting the approval of the rest of the board: the lack of consultation may be wrong, but it has nothing to do with the theft.)

The point is that the fact that the mother's body is directly involved in gestating a foetus gives her a golden vote on the matter. For sure: it might be admirable for the mother to consult the father, and in certain cases we might think less of her for not doing so – but there would be no overriding obligation. And while it is also possible for some to argue that the father's feelings would be a relevant consideration in the matter of assessing the permissibility of continuing with a termination of pregnancy – if he really wants the child to be born, mightn't this count against the maternal appeal to bodily integrity? – such a line would only stand a chance of being successful if (a) one is committed already to some kind of utilitarian calculus and (b) one can find some way convincingly to make the value of the father's feel-

[5] Actually, this is not the only problem: one peculiarity of the Ohio Bill is its insistence that positive approval must be forthcoming from the father, rather than simply saying that he may veto an abortion. This means that the father is taken to have a claim over the mother's body in all circumstances – even if, after the condom splits during a one-night stand, he disappears over the horizon never to be seen again. This does somewhat give the lie to the "joint project" defence of the Ohio Bill. On the other hand, this is a problem with the Bill – but it does not amount to a fatal flaw in the defence, which still works in principle, even if it turns out not to fit the Bill too well in practice.

ings commensurate with (let alone significant in comparison to) the value of the mother's bodily integrity. In other words, to the extent that they are equal partners in the parenthood project, it makes sense to say that each parent has an equal input in decisions about parenthood. But, as far as gestation is concerned, fathers will always be the junior partners; and it makes sense to say that their input will be the less.

How Glass Wombs Make a Difference

Be all this as it may, affairs are made more complicated when we take into account the possibility of ectogenesis and gestation *in vitro*. For if what makes the difference between a mother's and a father's say in whether or not to go ahead with a foeticidal procedure is the fact of gestation, it would follow that, were it possible to pare gestation from the question, the moral balance would shift back towards the centre of their relationship.

To some extent, such a paring is possible, and we are having to get to grips with the moral questions raised thereby very quickly already. For example, in cases of surrogacy, the moral authority of the commissioning mother over the continued life of the foetus might properly have to take second place to that of the gestatrix in some cases – for, while it is "her" foetus that is gestating, it is not her body that is gestating it. To this extent, the commissioning mother's moral position is roughly similar to that of the father, inasmuch as that her gamete is gestating inside another's body.

In other cases, when an embryo is stored outside of the mother's body, we may feel that the father's approval is a relevant consideration in decisions about whether that embryo should be implanted and brought to birth. Famously, it was a dispute between a woman and her ex-partner concerning whether she had the right to implant the embryos they had jointly created prior to her cancer treatment that led to the principle of paternal consent being confirmed by the British courts and upheld by the European Court of Human Rights. (In this case, the ECHR found that, under the European Convention of Human Rights' Article 8 protection of private life, that protection extended to respect "for both the decisions to become and not to become a parent" and noted further that the applicant, Ms Evans, "had not complained that she was in any way prevented from becoming a mother in a social, legal, or even physical sense, since there was no rule of domestic law or practice to stop her from adopting a child or even giving birth to a child originally created *in vitro* from donated gametes". The court ruled that there was no strong ground to accept that Ms Evans' "greater physical and emotional expenditure during the IVF process, and her subsequent infertility, entailed that her Article 8 rights should take precedence over [her ex-partner's]." Additionally, it was held that "[Ms Evans'] right to respect for the decision to become a parent in the genetic sense should [not] be accorded greater weight than [her ex-partner]'s right to respect for his decision not to have a genetically-related child with her"[5].)

Already, then, new reproductive technologies raise questions about the balance of authority over the continued life of a foetus or embryo that, not so long ago, would have been purely formal. As Shultz noted as far back as 1990, "[d]evelopments in reproductive technology have created new biological and social options that in turn challenge old assumptions and pose new dilemmas for legal doctrine and policy" [6].

At present, surrogate pregnancy represents the most radical possible severance between gestation and parenthood. Gestating a foetus in a manmade medium is, at present, not possible. But the impossibility of *in vitro* gestation is technical rather than conceptual. The barriers could be overcome within a matter of a few years granted a little effort and some judiciously-distributed funding. (Given that it is already possible and fairly frequent for an egg to be fertilised *in vitro*, and that we are capable of saving the lives of ever-more premature babies, we may imagine a time tolerably soon when the two lines meet, and when the whole process of foetal development could take place outside the body: a woman would go to the clinic to have a certain number of eggs harvested, have them fertilised with the chosen father's sperm, and then come back nine months later to pick up her child.[6])

A device capable of *in vitro* gestation might well be welcomed by several groups of people: women who desire children but who, due to illness or congenital abnormality, find themselves unable to sustain a pregnancy represent a prime example. The use of a "glass womb" might well turn out to be comparably desirable in certain cases of extremely premature labour or in cases when the mother has been injured and continuing with the pregnancy conventionally would threaten her survival: we could, in principle, simply continue with the pregnancy outside of the body. However, such a device would, we might expect, also be welcomed by those women who desire children but who are unwilling for some reason to go through the process of natural gestation. Finally, we ought not to forget the attraction of a glass womb to gay men (or straight men for whom natural conception is, for some reason, not available), who could choose their desired egg from an egg bank and have it fertilised with their sperm and gestated without the need to engage a surrogate – at least, not a walking, talking one.

Several questions present themselves in the light of such a possibility, though. One has to do with the right to terminate a pregnancy, and the robustness of any right that there may be to do such a thing. Another has to do with what the possibility of such interventions tells us about the "meaning" of motherhood and womanhood now, compared to what it implies about their possible future meaning.

Terminations and Glass Wombs

The factor that upsets pro-lifers about terminating a pregnancy is that such a process invariably implies foeticide. This is why inducing delivery when a pregnancy overruns is not ordinarily thought of as a termination, even though terminating the

---

6 Indeed, why nine months specifically? If, say, a year's gestation turns out to be in the baby's best interest, why not opt for that?

pregnancy is precisely what is intended. It is also why, in normal conversation, we can generally afford not to worry whether the word "abortion" refers to aborting the pregnancy or aborting the foetus' life, since the former implies the latter. But, were it possible to terminate a pregnancy without thereby ending the life of the foetus, we could expect a corresponding change in the nature of the "abortion debate". It would still be open to question whether the foetus has a right to life; but the chances are that at least some proponents of the right to terminate a pregnancy would find themselves having to clarify their position. After all, if your defence of abortion rests on an appeal to the bodily integrity of the mother, you might presently find yourself accepting the death of the foetus with a heavy heart, as something unavoidable but undesirable. The availability of ectogenesis, though, could alter this: we would find ourselves in a world in which a woman could terminate her pregnancy without thereby having to accept foeticide.

Moreover, were ectogenesis to become a practical possibility, then the right of the mother unilaterally to decide to terminate the pregnancy would not have to indicate the right of the mother unilaterally to decide to act foeticidally. For the possibility of *in vitro* gestation separates gestation from pregnancy; and while granting the mother final say over whether to continue with a pregnancy at present means having to concede that she also has the final say over whether the foetus lives, the potential use of a glass womb underlines that there is no necessity in the connection between the two aspects of her decision. She could perfectly well decide that she wanted to end the pregnancy without thereby having to end the foetus' gestation.

This being the case, though, one of the possible defences of the Ohio Bill considered a short time ago looks as though it might well be reinvigorated, to at least some extent. What gave the Bill its – admittedly limited – moral power was the thought that parenthood is a joint project; what stymied it was the thought that gestation is not. But, granted the possibility that the day may well come – and come fairly soon – when it is possible both to fertilise and then to gestate the egg *in vitro*, then the physical proximity between mother and the developing foetus becomes a null consideration: they need not even be on the same continent.

If this line of thought is weighty, it means that there is something to commend the claim that the male parent has a say in the future survival of the "ectobaby" equal to that of the female parent, because that future would represent the interests of both equally. Indeed, it also means that, if we think that parents do have a say in the future of the foetus, we would have no choice but to consult the father. After all, the function of the father is, biologically speaking, nothing more than to produce a gamete to be fused with another gamete and gestated somewhere else, and since the possibility of *in vitro* fertilisation and gestation represents the possibility of this function being fulfilled by both men and women, there is a sense in which parents of either sex would be, in effect, "fathers". (Granted, retrieving eggs from a woman is more difficult than is retrieving sperm from a man, but such a difference does not seem all that important in the present context.) The father would *have* to have a say in the future of the foetus if either parent did, because both would be "fathers".

There would be no "mother".[7] To exclude one parent simply because of his/ her sex would be straightforwardly sexist.

So, for the argument, let us imagine that we live in a world in which ectogenesis is a practical possibility: call this world "Ectoworld". To the extent that the Ohio Bill could be interpreted as representing a claim about parenthood being a joint project, then its analogue in EctOhio would perhaps have something going for it.

Imagine that a couple fertilises and gestates a foetus *in vitro*, but some reason presents itself to end its life. If foeticide is permissible, and if a foetus is the manifestation of a joint project of parenthood, then on Ectoworld neither parent would have the *de facto* authority to go ahead with the procedure without the approval of the other. A parent-to-be who no longer wanted to raise a child would be, presumably, in the same position as a parent who no longer wanted to be a parent in our present world – that is to say: committed.[8] Conversely, if foeticide is impermissible, neither would have the scope to approve it.

What about pregnancies *in vivo*, though? Granted, again, the joint project model of parenthood, in those cases when a mother wants to end the pregnancy but the father does not want the foetus to die, might he have a right of some sort to insist on ectogestation? After all, the shortcoming of the Ohio Bill was that it prioritised mere parenthood over the practicalities of gestation for the woman. But if the facts of maternal gestation could be put to one side, it would be unclear why the gestatrix would have to be a woman, and why either parent should retain a golden vote as a result. The parents' positions would be, we can assume, equal.

The obvious rejoinder to this line of thought is to point out that to gestate a foetus *in vitro* after having conceived it *in vivo* would require a medical procedure to retrieve it; and for a father to insist that his partner undergo such a procedure might be seen as no less – and maybe more – of an attack on her sovereignty over herself and her body than would be for him to veto a termination (since to vetoing a termination of pregnancy, unlike transposing a foetus, only preserves the *status quo ante*). But even this rejoinder is not straightforward; for in a situation in which the father wants the mother of his child to undergo a procedure to move a foetus when she wants the pregnancy over, we would be faced with a situation in which her alternative course of action would be to undergo a different, foeticidal, procedure. The choice is not between an operation and no operation; it is between two different kinds of operation, given that one is going to take place either way.

Ordinarily, of course, an agent's decision concerning whether to have medical procedure A, B, or neither, is that agent's alone. But that is because, ordinarily, other people are not involved; the problem with parenthood is just that others *are* involved and would seem therefore to have at least some right to representation in

---

7 Except maybe the machine, or the technician whose labour – pun fully intended – brings the child to birth.

8 The difference between the parent in this sort of example and the father in *Evans* is that bringing a child to birth in the latter case required a *further* procedure involving both patents' genetic material, implying a need to get both parents' consent; no such further procedure would be necessary in the former, ectogenetic, case.

respect of decision-making, even if that right is frequently only nugatory. If, though, the mother is going to have some kind of invasive procedure anyway, it would seem to be possible to ask whether the interests of the father (and the foetus itself) in the foetus' continued life might be such as to mean that they are entitled to some say in deciding which invasive procedure is chosen.

I shall take it as read that foeticidal procedures would be less invasive than transpositional procedures. If – implausibly – it turned out to be the case that the foeticidal procedure was the *more* invasive option, and if there were no independent reason to prefer that the foetus not survive (such as a disability sufficiently grave to mean that the child's life would be burdensome), there would be a clear incentive to opt for the foetus to be transplanted into a glass womb. The same would apply if the two procedures were of exactly equal invasiveness, since it seems reasonable to take as axiomatic the notion that (*ceteris paribus*) a world in which the foetus survives is better than one in which it doesn't.[9] But since her role as gestatrix would end whichever option was chosen, the mother would relinquish her automatic priority in deciding what happened to the foetus either way: take the baby from her belly, and her interests in the parental project and desires *vis-à-vis* the foetus are equal in weight to the father's. (That is to say: she would have an inalienable priority in questions of whether or not the foetus stayed within her; but on questions of what happened to the foetus thereafter, she would have no privileged position.) If, though (and as seems more likely), the foeticidal procedure were to prove *less* invasive than the transplant procedure, and if all else were equal, then the father would, quite properly, have no right to insist on the transplant. Whatever his entitlement in relation to the joint project of parenthood, he has no claim over the body of the mother.

Still – what about the pro-life rejoinder that all else is *not* equal, since a human life (at least after a given point) is of such value as to shift the balance? After all, situations in which a life is at stake are often situations in which we are entitled to break or bend otherwise-robust rules. Imagine that Smith and Jones work in a noisy factory; Smith is about to pull a lever that will result in Jones' death. Because Smith cannot hear our warnings, we might be entitled to throw a stone or small bolt at him to distract him and stop him, even though we would not dream of throwing anything at him in any other circumstance. And in situations when something as important as a human foetus is at stake, we might think that there are analogous exceptions to general rules about non-interference in others' lives: even though we would not dream of imposing a procedure on a person the rest of the time, in this case, we are entitled – maybe even obliged – to do just that.

And one does not have to be particularly pro-life to accept a recognisably similar conclusion: pro-choicers might well accept that medical invasiveness and appeals to sovereignty could turn out to be moot given the intuition that, if we can save a life

9 I am assuming that, were the foetus gestated *in vitro*, there would be no obligation on the parents to raise it after its "birth": they would be entitled to offer it for adoption and need never see it again. Whatever we think about parents willing never to see their offspring again, it has nothing to do with the ethical problems raised by ectogenesis.

thereby, there are certain sacrifices that we might be obliged to make. If Thomson's violinist will inevitably die when we unhook him, and this is an unavoidable part of our exercising our right to bodily integrity, then his death might be something regrettable without being blameable. But if we could unhook him from ourselves and have him hooked to a suddenly-available ectokidney, then it might well strike us as unforgivable that we consign him to oblivion anyway, and we might think that an observer has a right to force us not to do so. For many people, this would seem the best of all possible worlds.

Indeed, even if one thinks that the foetus has no rights at all, one might still think that there is something blameable about causing the death of something that need not die. For Kantians, the appeal here may be to the kind of indirect duty described in the *Doctrine of Virtue* [7]; for virtue theorists, the appeal may be to the dishonourable pusillanimity and arbitrariness shown; for utilitarians, the appeal could rest on the claim that, in killing the foetus, we are damming a source of possible future welfare. Overall, and whatever our wider commitments, the easier it is to move the pregnancy into a glass womb, the more plausible might seem one of two claims: either that the father has a *right* to insist on such a move over any procedure that would result in foetal death (based on his joint stock in the parental project), or that he has a *duty* to do so (based on the obligations we all have to protect human life).

Now, one possible worry here might be that such a suggestion simply entrenches a male power over the woman. In fact, though, it does no such thing – whatever power relations there are would have nothing to do with sex. We could imagine, for example, some technology that allowed a foetus to be created using genetic material from two eggs and no sperm, and then implanted. Thus a lesbian couple could have a child that is related to each of them to exactly the same extent as it would be related to heterosexual parents. Exactly the same considerations about the end of the pregnancy would obtain. For as long as both parents, be they of the same or different sexes, have some interest in the survival of the foetus *qua* foetus – and they do – then it would remain true that the gestatrix would have sovereignty over her body, but not necessarily sovereignty over the life inside it. Similarly, *if* it is morally permissible for the non-gestating parent to request or insist on a non-foeticidal end to the pregnancy – and I have made a point of not coming down one way or the other on this matter – then it is nothing to do with that parent's sex: it's just a facet of the relationship that exists between any two agents. Even if these lines of argument are not, in the end, sufficient to dissolve a claim that foeticide is permissible, they do at least indicate that such a claim would not be guaranteed an easy run.

An assessment of the permissibility of foeticide given the possibility of ectogenesis, and how we might evaluate the sacrifices that it is reasonable to demand of an agent – a mother in this case – to avoid another's death, is beyond the scope of this paper. However, such questions are, in their way, fairly workaday: they are already asked in relation to the abortion debate. What I hope to have established so far, though, is that the manner in which we go about answering those questions in the present world is often informed by the role of the woman-as-gestatrix and her corresponding physical proximity to the foetus. Alter that role, and the shape of the abortion debate – specifically, in relation to extent of the mother's authority over the future of her pregnancy – alters as well. Implicitly, too, as the shape of the moral debate about abortion is altered by technological possibility, then so too is the shape of its legislative correlate: lawmakers perhaps would do well to consider how they should respond to possible innovations in reproductive technology well before those possibilities are realised.

Ectogenesis and Gender

I mentioned a little while ago that ectogenesis allows members of either sex to play the father. This was a touch mischievous, and intended to bring to the fore a worry that might be articulated about the technology.

For some, ectogenesis is a cause for optimism. In particular, and in addition to the possibility that would offer the chance to have children to at least some women who would not otherwise be able to do so, it offers women a degree of reproductive freedom: it promises liberation from the medical risks of pregnancy and from the straightforward inconvenience. Fathers, hitherto, have had a pretty easy time of reproduction: as a bonus, ectogenesis offers to women the chance to share in that ease equally. The inconvenience of harvesting eggs is small beer compared to the inconvenience of *in vivo* gestation.

But the idea that women could play the father could well raise worries, too. For some theorists, the thought may be that even if the advent of the glass womb does bring equality and liberation of a sort, it is equality and liberation of a male sort. Perhaps, in Ectoworld, "female" or "woman" would become entirely superfluous terms, so alienated would women have become from their bodies' capacities and functions. Ectoworld would be a place with two kinds of men: spear-men and distaff-men, the only real difference being the ease with which gametes could be retrieved from differently-shaped genitalia. Sally Sheldon, for example (admittedly in a different context), articulates the point that reproduction does seem to be a peculiarly – and visibly – female activity. It is the woman's body which carries a pregnancy, which provides the necessary warmth and nutrition to sustain the foetus, and which swells to accommodate its growth. By comparison, the man's involvement seems insignificant, amounting to nothing more than the ejaculation of a small quantity of seminal fluid [8].

Remove the warming, feeding and swelling from the female body, such that it, too, does nothing save providing a gamete, the thought goes, and we end up removing its femininity.

To see this as liberation is, arguably, to see the male role as the ideal to which any sane agent would aspire. The brute facts about being a woman and, in particular, about being a gestatrix, on this account, would have been presented as problems to be overcome. Indeed, such "liberation" seen through this prism would precisely *not* give women reproductive equality: it would give them reproductive sameness with men, while reinforcing the notion that the female reproductive role is, by its very nature, very much the inferior to the male.

Such a deconstructive line of argument may or may not be attractive: for many, it may be far too essentialist and teleological – too intent on telling agents what they are for and perhaps reducing them to gestatrices, it seems (to those who adhere to the egalitarianism Wollstonecraft and her heirs) simply to misfire for women. I shall leave aside debates about the importance of difference here and allow, for the nonce, that the deconstructive account is correct, and that ectogenesis is something that erodes the concept of what it is to be a woman. Let's imagine, too, that we live in a world in which everyone has the choice of whether to reproduce naturally, or whether to take a trip to the local laboratory and let someone else do most of the work. And let's imagine, finally, that everyone happens to decide, from a wholly neutral starting point, that a trip to the laboratory would be far better. This world, if the worriers are correct, would be a world in which womanhood is threatened, if not already completely obsolete.

Would this possible world be worse than ours in any way? I am tempted to think that, whatever it may lack compared to ours, it would not. My lead here comes from considerations articulated by Anthony Ellis in respect of minority languages. In the first place, a central theme of the deconstructive line is that of difference and parity of regard for different characteristics – it is not committed to saying that x is particularly good or better than y, but to questioning the foundations of such ordering. Thus, presumably, we can say that, from a god's-eye position, the male and the female both have their own different and unique virtues. Very well: this would presumably mean, too, that for an agent to abandon (archetypally female) natural gestation in favour of (non-archetypally-female) ectogenesis is to choose a role that is neither better nor worse: just different.[10] The price paid in adopting any role is the opportunity to adopt any other role (for the time being), but that is not a bad thing: it is just a fact of only having one life to live, and the agent would have been able to choose between competing good things. Now, Ellis suggests that

> [i]t is no doubt a good thing to know Welsh and to be able to read its literature. But I assume that this benefit is offset by the fact that learning Welsh involves not learning something else that it would be beneficial to learn [9].

– which, on the face of it, could be rewritten to convey the idea that

> [i]t is no doubt a good thing to live one's life according to the feminine model. But I assume that this benefit is offset by the fact that leading one's life according to it involves not leading it in some other manner that would be good.

For sure, being *forced* to speak English rather than Welsh, or to conform to a masculine rather than feminine paradigm, would be a bad thing – but this is not because that which one is forced to do is bad, but simply because one is forced to do it. As

[10] And, while we're at it, why not imagine men who, with a little surgery, became gestators? They would not be taking on a lesser role: they would be taking on a *different* one. While some may imagine that this would facilitate yet another male intrusion into female life, it would do so no more than would *in vitro* gestation facilitate female intrusion into male life. If the two are different but equal, there should be no problem with this: we would be dealing with *adopting* a role, not intruding on it.

long as one can seek to live one's life and to flourish according to one's beliefs about the nature of that life, there oughtn't to be a problem.

Would it matter if everyone chose to abandon the feminine role? Again, I think not: again, Ellis provides my lead.

> The world would have been an immeasurably poorer place if Latin had disappeared without trace; but I do not see that it would have been an immensely better place if some isolated community had retained Latin as a living language [9].

This seems correct. The riches of Latin, in other words, ought not to be fetishised: it is, in the final analysis, just another language, with no inherent virtue denied to any other. The same would presumably apply to gender paradigms: there is no inherent virtue in preserving diversity for its own sake, and, if one paradigm dies a natural death, then that is just too bad. So it goes. On the other hand, it is not clear that any paradigm – linguistic or gendered – *would* disappear without trace anyway: we learn Latin at school, and it lives on in Romanian and most of the languages of Western Europe; *mutatis mutandis*, we might learn about "womanhood" as something that our ancestors used to do, and we might live according to the norms of its various recognisably descendant paradigms. Nor is it clear that, even if it did vanish utterly, we would notice, since we would all be far too busy speaking some other language or playing some other role (again: noone is bothered by the fact that whatever language was spoken by our prehistoric ancestors has gone for good). There would be no chasm needing to be filled.

I do not think that femininity is threatened by ectogenesis – I would want some evidence before I believed it – but, even if I am wrong, the most that we would lose is a certain way of living a life. And since, *ex hypothesi*, this way would be neither better nor worse than any other way, and since agents would not be left bereft of any way at all to live their life, I cannot see how even that loss ought to provoke much of a worry.

Acknowledgments This paper has benefited from discussions with Amel Alghrani and Becki Bennett, to whom thanks are due.

References

1. House Bill 287; available via http://www.legislature.state.oh.us/bills.cfm?ID=127_HB_287
2. Mill, John Stuart. 1998. On Liberty. repr. in Mill, John Stuart (ed.). On Liberty and Other Essays. Oxford: Oxford University Press
3. Kant, Immanuel. 1993. Grounding for the Metaphysics of Morals. Indianapolis: Hackett
4. Thomson, Judith Jarvis. 1971. A Defence of Abortion. Philosophy and Public Affairs 1: 47–66
5. European Court of Human Rights. 2007. Information note 96
6. Shultz, Marjorie Maguire. 1990. Reproductive Technology and Intent-Based Parenthood: An Opportunity for Gender Neutrality. Wisconsin Law Review 1990: 297–398
7. Kant, Immanuel. The Doctrine of Virtue. 1998. The Metaphysics of Morals. Cambridge: Cambridge University Press
8. Sheldon, Sally. 1999. ReConceiving Masculinity: Imagining Men's Reproductive Bodies in Law. Journal of Law and Society 26: 129–149
9. Ellis, Anthony. 2005. Minority Rights and the Preservation of Languages. Philosophy 80: 199–217

Chapter 16
A Survey of People's Attitude Towards the Artificial Womb and Ectogenesis in Israel

Frida Simonstein and Michal Mashiach-Eizenberg

Introduction

As suggested elsewhere [1] and also in this volume (see Chapter 13), the artificial womb (AW) and ectogenesis – a child brought to term outside a biological womb, might become a reality sooner than we may think. Novel research in disparate areas (such as neonatal care, assisted reproduction, embryology, fetal surgery, computer science and the human genome project) are converging to this end. Society is also pressing in this direction. Society, at large, aims at saving very premature newborns, and demands better outcomes in IVF. Because of this pressure, both sides of the gestation process, its beginning and its end (conception and birth), are presently experiencing a massive research effort. Indeed, while there is sill a huge gap between the first stages of gestation (by IVF) and the 22nd week of gestation (inside the womb), plausibly, this gap will be eventually overcome.

Ectogenesis has been addressed in many scientific – philosophical writing [2–4] as well as in popular media. And while some medical sources avoid today discussion on this topic, others choose to address the advent of the artificial womb as a matter of fact [5]. The artificial womb, however, is not exactly welcomed: most countries ban research on human embryos beyond day 14th. Canada has explicitly prohibited any research designed to add to the knowledge of ectogenesis [6].

Still, since prematurely born babies spend less time in a woman's womb, the question of why and for how long an embryo 'should' be required to be in a woman's womb (nine months; six; five? two days?) cannot be easily answered. Moreover, if proved one day to be safe, ectogenesis, might be a solution to avoid the need of a surrogate for women, who want a biological baby, but lack a womb (for any reason); and possibly also for homosexual male couples. It might also appeal to women, who may need IVF to reproduce; women may save time, pain, depression and endless frustration (when embryos fail to implant for instance).

F. Simonstein (✉)
Department of Health Systems Management, Yezreel Valley College, Yezreel Valley, Israel
e-mail: fridas@yvc.ac.il

F. Simonstein (ed.), *Reprogen-ethics and the Future of Gender*, International Library of Ethics, Law, and the New Medicine 43, DOI 10.1007/978-90-481-2475-6_17,

In general, the AW may prevent damage to women's health since it may avoid short and/or long term negative effects which may appear during or after a pregnancy [7–9]. For although a pregnancy is not considered an illness [10], this is not exactly a healthy enterprise either. And without proper medical care, it may be deadly [11]. Even more contentious, though, as Tuija Takala points out in Chapter 14, ectogenesis may finally achieve women's equality; since a woman who wants a child might not be forced necessarily to slow down her career because of a pregnancy. Although, generally, gender inequality results mainly because of child care and a vast range of other gender related issues, and not necessarily because of the pregnancy, its roots may be found in the fact that women (unlike men) are those who bear the child and must slow down. For some women, the option to *really* choose to bear the child (or not) in order to become a parent, could be of value.

These issues are highly controversial; yet little is known about the public's views on these topics. This study, therefore, aims at starting to fill this gap: Its purpose was to explore Israeli people's attitudes towards the advent of the AWs. Awareness of people's views on this topic should be useful for further discussion and appropriate legislation. Although ectogenesis and the artificial womb are not the same, (ectogenesis is the process and the artificial womb is the tool), in this paper, we use these expressions interchangeably.

Research Methods

The study included 216 subjects aged over 21 who answered a structured self-report questionnaire. The questionnaire was divided in two sections: The first part addressed personal data of the respondent (such as gender; age; family status; number of children, profession). The second part targeted the views of the respondent towards the AW; this part included 12 statements ranked on a Likert scale between 1 – strongly disagree, and 5 – strongly agree. Half of the statements were negative and the other half positive. The 'position score' was calculated as the average of all the answers by the respondent (after the appropriate reversal of the scale for the negative statements). A score close to one determined a strong negative position of the respondent – strongly against the AW. A score near to five expressed a positive stand – i.e. strongly agree with this development. The internal consistency (Cronbach's alpha) of the position score was 0.899. The sample in this survey was obtained by a snowball method (i.e. referrals of respondents, which generated additional subjects). One obvious limitation of this way of sampling is that it may affect the representative nature of the sample. However, although not representative (there are fewer men), the sample in this study does include different sections of the Israeli population (Table 16.1).

Table 16.1 shows that the percentage of women in this survey was three times higher than the percentage of men, which may suggest that men might be less responsive to this questionnaire than women. The majority of the respondents in this survey were aged 31–51; but nearly a third of the sample were younger than 31; the

Table 16.1 Personal characteristics of the subjects

	Response (N)	Percents (%)	Mean±(SD)
Gender			
man	51	24	
woman	165	76	
Age			35.8±(9.94)
21–30	68	31.6	
31–50	131	61	
51–and older	16	7.4	
Family condition			
Single	69	31.9	
Married	138	63.9	
Divorced or widower	9	4.2	
Children			1.61±(1.54)
0	82	38.7	
1–3	108	51.0	
4 or more	22	10.3	
Religion			
Jewish	156	72.6	
Christian, Moslem or Druze	68	27.4	
Religious level			
Religious	20	9.3	
Traditional	59	27.3	
Secular	137	63.4	
Origin			
Israeli born	155	72	
Immigrant	59	28	
Education			
High school	22	10	
Non-academic professional training	97	45	
Academic	96	45	

sample included also a smaller proportion of people older than 51. Most subjects were married; but nearly a third was single and few were divorced or widowed. The majority of respondents had 1–3 children; a small proportion had more than four children and a third had none. Most of the subjects in this study were Jews; but there was a good representation of non-Jews (Christians, Moslems or Druze). While most of the respondents were secular, there was also a good presence of the so called 'traditional' and a smaller percentage were religious in various degrees. The majority was born in Israel; but there was also a fair representation of immigrants. An equal part of the respondents had either academic education or some non-academic professional training, while a smaller fraction had just 12 years of education.

Results

While people's attitude towards the AW was explored through randomly ordered statements in the questionnaire for better clarity in this report the responses were grouped in four distinctive categories. The first group contains statements reflecting a general position towards the AW; the second relates to statements about possible uses of the artificial womb. The third category includes possible uses of the artificial womb specifically related to women's reproductive chores. In addition, the questionnaire examined to which extent in two given situations the respondent would actually use the AW: in order to save their own fetus and/or to avoid a pregnancy (Table 16.2).

Table 16.2 shows that just 10.2 percent of all the respondents thought that the artificial womb should not be developed under any condition; by contrast, 76.8 percent of all respondents were against this expression. While 47.2 of all respondents agreed that the development of a fetus in AWs is against human nature, a majority of 57.9 percent disagreed with the phrase stating that the development of a fetus in AW is against human dignity. Sixty five percent of all respondents agreed that it is important to develop the artificial womb to save the life of premature fetuses. Fifty three percent agreed with the expression that it is important to develop the artificial womb for research to improve the effectiveness of IVF. Nearly forty three percent of the respondents agreed that it is important to develop the artificial womb to avoid the need of surrogates; and notably, a large majority of 74.5 percent thought that it is important to develop the AW for women, who do not have a womb. The majority of respondents, 71.3 percent, thought they would use an AW in order to save their own fetus. Opposing to this relatively large percentage of positive views towards the AW – when women's chores in reproduction were at the center of the statements, people's views became less receptive: nearly 51 percent of all respondents thought that women should not be given the opportunity to freely choose to use the artificial womb. Fifty nine percent thought it is unimportant to *develop* the AW so women could choose the way of giving birth. And 67.9 percent thought it is not important to develop the artificial womb to ease women's reproductive tasks. Consistently, a large majority, 87 percent of the respondents, thought they would not use the AW to avoid a pregnancy. We further analyzed the link between socio-demographic conditions and people's position towards the advent of AW An analysis of the correlation between socio-demographic background and people's views is shown in Table 16.3

Table 16.3 shows that men and singles of both sexes had a significantly more positive approach towards the AW; there was a negative correlation between the number of children and the attitude towards the AW, i.e. respondents with fewer children had more positive attitudes to this development. Jewish respondents were significantly more positive towards the development of AW, than non-Jewish respondents. There was also a negative correlation between the level of reported religiosity and the acceptance of AW, i.e. secular people were more positive towards the artificial womb. Finally, there was a significant difference between the mean attitudes towards the AW according to educational level, thus, respondents holding an academic degree were more positive to this development.

Table 16.2 Attitudes towards the artificial womb (%) (N = 216)

Item		Proportions (%) of responses			
		Disagree	Partly agree	Agree	Missing
General position towards the artificial womb	The artificial womb should not be developed under any condition	76.8	11.6	10.2	1.4
	The development of a fetus in AW is against human nature	35.2	16.7	47.2	0.9
	The development of a fetus in AW is against human dignity	57.9	21.3	19.9	0.9
Possible uses of the artificial womb (It is important to develop the artificial womb for:)	Saving premature fetuses	18.5	15.7	65.3	0.5
	Research to improve IVF effectiveness	22.2	22.7	53.2	1.9
	Avoiding the need of surrogates	36.6	19.9	42.6	0.9
	Women who do not have a womb	9.7	15.3	74.5	0.5
Possible uses of the artificial womb (related to women's reproductive chores)	Women should not be allowed to have the choice of using the artificial womb if they want to	36.7	11.1	50.9	1.4
	It is important to develop the artificial womb to make women's tasks in society easier	67.9	18.9	13.0	0.9
	It is important to develop the artificial womb to enable women choosing the way of giving birth	59.3	13.0	26.4	1.4
Possible use of AW by the respondent	I would use the artificial womb (for myself or for my partner) to avoid a pregnancy	86.5	3.2	7.8	2.3
	I would use the artificial womb to save my fetus if it is danger	15.8	13.0	71.3	0.0

Table 16.3 Relationship between socio-demographic background and attitude towards the AW (mean attitude index)

Variable		N	Mean	Standard Deviation	t test	Correlation
Gender	Woman	165	3.00	0.85	t(214) = 4.87**	
	Man	51	3.67	0.83		
Family condition	Single	69	3.38	0.90	t(205) = 2.56*	
	Married	138	3.05	0.88		
Religion	Jewish	156	3.31	0.85	t(205) = 4.34**	
	Not Jewish	59	2.75	0.82		
Education	Academic	96	3.30	0.91	t(213) = 2.00*	
	High-school	119	3.06	0.86		
Age		215				r = 0.032
Children		215				r = −0.278**
Religious level		216				r = −0.364**

Significance level: *P<0.05, ** P<0.001

Discussion and Conclusions

In a recent paper, following Shulamith Firestone's stand on ectogenesis [12], Anna Smajdor has suggested that instead of putting the burden on women to have children at times that suit societal, rather than women's individual interests, society could advance 'technical alternatives', such as artificial wombs [13]. Smajdor, challenges her readers with this question:

> You, the reader, from behind the veil of ignorance (Smajdor refers here to John Rawls' thought experiment)[13] are asked whether you would prefer to be born into society A, where women bear all the risks and burdens of gestation and childbirth, as they do now, or society B, where ectogenesis has been perfected and is routinely used. You do not know whether you will be born as a man or a woman. Which would you choose? [14].

The present study echoes Smajdor's hypothetical question. Yet, contrary to her implicit hope that people would choose society B, the large majority of people in this study chose in fact to be born into society A. Moreover, more women than men in this survey would prefer to bear all the risks, and burdens of gestation and childbirth traditionally associated with the nature of womanhood.

Nevertheless, the present study suggests that generally, people's attitude towards AWs is not as negative, as it could be expected. While people in this survey were markedly against using AW for ectogenesis, the majority did not think that the artificial womb should not be developed under *any* circumstance. Most of the respondents agreed that the AW is not natural to human beings; however, a large proportion of the respondents disagreed with the expression that the AW is against human dignity. Remarkably, three out of four people in this survey found acceptable the idea that AWs could resolve the problem of childlessness for women who do not have a womb. Respondents also attributed significant value to the development of AW to save the lives of premature fetuses (and for IVF research). Most tellingly perhaps, the majority of people in this survey thought that they would use AWs to save the life of their own fetus, if it happened to be in danger. This position reflects the societal pressure towards the development of AW.

Sharply contrasting with this position, however, when the idea of easing women's 'natural' roles in reproduction was at the center of the statements, the AW became unacceptable. The majority of the respondents disagreed with the idea that it may be important to develop AWs to give women another option for having children; and disagreed with the notion that women could be allowed to freely choose AWs. Most people also did not accept the statement that it is important to develop AWs to make women's lives easier. In short, neither the reduction of women's reproductive chores, nor women's option to have such a real choice on reproductive matters was at the center of people's acceptance of AWs. Consistently, most people in this survey thought they would not use AW to avoid a pregnancy.

Notably, the mean attitude of *men* in this survey was significantly more positive towards AWs (although the number of men in this study was smaller). Further analysis of gendered positions revealed also that the percentage of women who approved the idea that women's reproductive chores could be alleviated by AWs was lower than the percentage of men. While 87 percent of the women in this survey (N = 165) agreed with the statement that pregnancy and childbirth are not healthy for women's body, 92 percent of the women thought that a pregnancy is not a disease. Eighty-nine percent agreed that pregnancy and childbirth are important experiences for womanhood.

We also asked the women who already had children (N = 105) to mark in a scale of one to ten the degree of difficulty they experienced during the pregnancy and in childbirth (1-very easy; 10-very difficult). Thirty nine percent of the women who had had children ranked the experience of a pregnancy as very difficult (8–10 in the scale) and 44 percent found very difficult the experience of childbirth. We did not find, however, a significant link between the general position towards the AW and the difficulty of the pregnancy and childbirth. Although, there was a positive correlation between the difficulty that a woman felt during the pregnancy and her willingness to use the AW to avoid a pregnancy ($p<0.05$).

The fact that women were significantly less receptive of AWs in this study is hardly surprising; since in Israel (because of religious and demographic reasons) there is a strong societal conditioning of girls towards motherhood. Israeli girls (regardless of religion) learn from early childhood that infertility is a curse [15]. Even secular, well educated and wealthy women in the Israeli society, believe that having children is their duty [16]. Correspondingly, the majority of the women in this survey agreed with the statement that pregnancy and birth are key experiences for a woman. Interestingly, however, this endorsement of womanhood was negatively correlated with age. Thus, older women disagreed more with this statement; which suggests that life experience, beyond the reproductive years, plays a role in women's perspective over womanhood.

Although most people in this survey were against using AWs to avoid a pregnancy, there was, nevertheless, a positive correlation between the general attitude towards the development of AWs and the level of secularity and education of the respondent. Thus in this study, secular and well-educated people were more receptive of AWs. This suggests that positive developments in this technology could enhance people's receptiveness of AW in Israel.

It is difficult to predict how people would actually react when AWs technology becomes fully developed and safe [17]. For instance, as recalled in Chapter 13, there was a strong aversion towards the idea of creating a child using IVF until the first 'tube-baby' was successfully born [18]. Since then, more than three million babies have been born by means of IVF [19]. Nevertheless, acknowledging and respecting population views is essential. Peter Singer, for example, suggested that AWs could resolve the outcry against abortions (i.e. aborted fetuses could continue their lives in AWs) [20]. Contrary to Singer's idea, however, a study targeting women who had abortions revealed that women who wish to terminate a pregnancy do not want, necessarily, to have their aborted fetuses kept alive [21].

In the present survey, just a small fraction of the respondents was willing to use AW to avoid a pregnancy; however, this could amount to a significant number of people in the population. This study suggests that further research addressing the population's views (in Israel and in other parts of the globe) on the advent of AWs is essential. We believe that additional study, in-depth analysis and a broader discussion of this topic is necessary in order to develop appropriate legislation.

References

1. See, Simonstein, F. Artificial reproduction technologies – all the way to the artificial womb? *Philosophy of Medicine and Health Care* **9**, 359–365 (2006).
2. Singer, P. & Wells, D. *The Reproduction Revolution: New ways to Making Babies*. Oxford: Oxford University Press (1984).
3. Harris, J. *Clones, Genes and Immortality*. Oxford: Oxford University Press (1998).
4. Strong, C. Overview: a Framework for Reproductive Ethics. In, *Ethical issues in Maternal-fetal Medicine.* Ed. Dona L. Dickenson. Cambridge: Cambridge University Press (2002).
5. Ber, R. Ethical Issues in gestational Surrogacy. *Theoretical Medicine and Bioethics*. **21**, 153–169 (2000).
6. Final Report of the Royal Commission on New Reproductive Technologies. Minister of Government Services, Canada. Ottawa: Canada Communications Group. (1993).
7. Ayers, S., Eagle, A., & Waring, H. The effects of childbirth-related post-traumatic stress disorder on women and their relationships: A qualitative study Psychology. *Health and Medicine*. **11**, 389–398 (2006).
8. Russell, R., Dundas, R., & Reynolds, F. Long term backache after childbirth: prospective search for causative factors. *BMJ* **312**, 1384–1388 (1996).
9. Altman, D. et al. Risk of Urinary Incontinence After Childbirth A 10-Year Prospective Cohort Study. *Obstetrics & Gynecology* **108**, 873–878 (2006).
10. Cook, R., Dickens, B.M., & Fathalla, M.F. *Reproductive Health and Human Rights*. New York: Oxford University Press (2003).
11. http://www.who.int/pmnch/en/
12. Firestone, S. The Dialectic of Sex: The Case for Feminist Revolution (1970. New York: Farrar, Straus and Giroux, 3rd ed.) (2003).
13. Smajdor, A. The Moral Imperative for Ectogenesis. *Cambridge Quarterly of Healthcar Ethics* **16**, 336–345 (2007).

14. Rawls, J. *A Theory of Justice*. Cambridge: Cambridge University Press (1971).
15. Kahn, S.M. *Reproducing Jews. A Cultural Account of Assisted Conception in Israel Durham.* North Carolina: Duke University Press (2000).
16. Remennick, L. Childless in the Land of Imperative Motherhood: Stigma and Coping Among Infertile Israeli Women. *Sex roles* **43**, 821–824 (2000).
17. 'A Report of the President's Council on Bioethics. 2004 'Reproduction and Responsibility', The Regulation of New Biotechnologies. Washington, D.C. http://www.bioethics.gov
18. Marantz, H. R. *Pandora's Baby.* New York: Houghton (2004).
19. Three million IVF babies born worldwide, Progress Educational Trust. http://www.ivf.net/ivf/index.php?page=out&id=2105.
20. Singer, P. *The Reproduction Revolution: New Ways of Making Babies*. Oxford: Oxford University Press, Peter Singer (1983).
21. Cannold, L. *The Abortion Myth: Feminism, Morality, and the Hard Choices Women Make.* Hannover: NH, University Press of New England (2000).

Index

FSC
www.fsc.org
MIX
Papier | Fördert
gute Waldnutzung
FSC® C083411

Kolibri 360 GmbH
Ferdinand-Jühlke-Straße 7
99095 Erfurt, Deutschland
produktsicherheit@kolibri360.de